Security and Privacy in the Internet of Things

Contents

Security and Privacy in the Internet of Things

Edited by
Syed Rameem Zahra
Mohammad Ahsan Chishti

CRC Press
Taylor & Francis Group
Boca Raton London New York

CRC Press is an imprint of the
Taylor & Francis Group, an **informa** business

A CHAPMAN & HALL BOOK

First edition published 2021
by CRC Press
6000 Broken Sound Parkway NW, Suite 300, Boca Raton, FL 33487-2742

and by CRC Press
2 Park Square, Milton Park, Abingdon, Oxon, OX14 4RN

© 2021 Taylor & Francis Group, LLC

CRC Press is an imprint of Taylor & Francis Group, LLC

ISBN: 978-0-367-85994-7 (hbk)
ISBN: 978-0-367-64908-1 (pbk)
ISBN: 978-1-003-01630-4 (ebk)

Typeset in Palatino
by SPi Global, India

Abbreviations

CNN	Convolutional Neural Network
DoS	Denial of Service
DDoS	Distributed Denial of Service
DNN	Deep Neural Network
IDS	Intrusion Detection System
IoT	Internet of Things
SPoF	Single Point of Failure
SVM	Support Vector Machine
RNN	Recurrent Neural Network

Preface

Recent technological advancements accompanied by the rapid proliferation of smart devices and high speed networks have helped IoT to emerge as a field of incredible impact, growth and potential. IoT maneuvers connection among different aspects of people's lives, interconnecting homes, offices, cars, industries, etc. However, this mushrooming of IoT devices with limited storage, compute and network capacities and the open nature of internet connectivity puts a considerable load on the shoulders of security professionals; wide range of IoT devices and applications are vulnerable to nefarious attacks by hackers. Risks to security and privacy may significantly diminish end-user's confidence in IoT and therefore impede its full realization.

This book breaks down the misconception that a back-up can secure individuals and organizations from getting hacked by identifying the major security flaws in the IoT environment. This book looks into all the security aspects of IoT starting from physical level to the highest application level. It explores the challenge of designing robust solutions to deal with various security and privacy threats in IoT. Moreover, a hands-on about the basic concepts of various open-source tools in the domain of IoT security will be discussed in the book. The book is a good way for the reader to pick up interesting directions of research that are currently being explored and also hints at directions that could take additional investigation.

The following topics are explored in this book:

- Intrinsic security vulnerabilities of IoT devices.
- Attacks launched on the IoT in recent past.
- Loopholes in IoT security services.
- Physical layer security of IoT.
- Principal aspects of IoT security.
- Blockchain based IoT architectures.
- Biometric authentication in IoT.
- Technologies and protocols from their security point of views.
- Blockchain and machine learning, fuzzy logic and neural network-based solutions to the discussed issues.
- Open-source tools for IoT.
- Use cases demonstrating the S&P concerns in various smart areas.
- Loopholes in the data privacy rights caused by the use of virtual assistants.

Editors

Syed Rameem Zahra is currently pursuing her Doctor of Philosophy (Ph.D.) from National Institute of Technology Srinagar. She has earned the Bachelor of Technology (B. Tech.) degree in computer science and engineering (CSE) from the University of Kashmir, Srinagar, India, and the M.Tech. degree in computer science and engineering from SMVDU, Katra, India. She is a gold medalist at both graduate and post graduate levels. She has published more than ten research articles in reputed journals (indexed by Scopus and SCI) and conferences (indexed by Scopus). She has been facilitated with a *Certificate of Appreciation* by Ministry of Human Resource Development (Department of Higher Education), Government of India for her academic excellence. Her research areas include the Database systems, Wireless Sensor Networks (WSN), Vehicular Ad hoc Networks (VANET), and Internet of Things (IoT) Security.

Dr. Mohammad Ahsan Chishti has earned his Doctor of Philosophy (Ph.D.) from National Institute of Technology Srinagar. He has earned Bachelor of Engineering (B.E.) and M.S. in computer and information engineering (MSCIE) from International Islamic University Malaysia. Presently he is Dean, School of Engineering & Technology, and Associate Professor in the Department of Information Technology, Central University of Kashmir. He has more than 70 research publications to his credit and 13 patents with two granted International Patents. He has successfully completed Department of Science & Technology sponsored research project. He has been awarded "*IEI Young Engineers Award 2015–2016*" in the field of Computer Engineering for the year 2015-16 by the Institution of Engineers (India) and "*Young Scientist Award 2009–2010*" from Department of Science & Technology, Government of Jammu and Kashmir for the year 2009–2010. He has been Executive Member, Institute of Engineers India (IEI) of J&K Chapter. He has been guiding Phd, Master of Technology (M.Tech) and Bachelor of Technology (B. Tech) research projects and his research area includes IoT, Computer Networks, Next Generation Networks, MPLS and Technology Roadmapping. He is Senior Member-Institute of Electrical and Electronics Engineers (IEEE) and Member of several other societies like IEI, CSI, IETE apart from being member of other technical societies.

Contributors

Mohammad Ahsan Chishti
Department of Information Technology
Central University of Kashmir
Tulmulla, Ganderbal, India

Aaqib Bulla
Communication Control and Learning Lab
Department of Electronics &
    Communication Engineering
National Institute of Technology
Srinagar, India

Fernando Nobre Cavalcante
State University of Campinas – UNICAMP
Brazil

Farah S. Choudhary
The Business School
University of Jammu
India

Anam Iqbal
Department of Computer Science and
    Engineering
National Institute of Technology
Srinagar, India

Rashmi Jain
Rajiv Gandhi College of Engineering and
    Research
Nagpur, India

Baijnath Kaushik
Department of Computer Science and
    Engineering
Shri Mata Vaishno Devi University
Katra, India

Mohammad Khalid Pandit
AI & ML Group
Islamic University of Science and
    Technology
Awantipora, India

Ab Rouf Khan
Department of Computer Science and
    Engineering
National Institute of Technology
Srinagar, India

Sahil Koul
The Business School
University of Jammu
India

Ambika N
Department of Computer Applications
SSMRV College
Bangalore, India

Salim Qureshi
Department of Computer Science and
    Engineering
Model Institute of Engineering and
    Technology
Jammu, India

Mohd Rafi Lone
Department of Electronics &
    Communication Engineering
Baba Ghulam Shah Badshah University
Rajouri, India

Faisal Rasheed Lone
Department of Information Technology
    and Engineering
Baba Ghulam Shah Badshah University
Rajouri, India

Shahid M. Shah
Communication Control and Learning Lab
Department of Electronics &
    Communication Engineering
National Institute of Technology
Srinagar, India

Sparsh Sharma
Department of Computer Science and
    Engineering
Baba Ghulam Shah Badshah University
Rajouri, India

Surbhi Sharma
Department of Computer Science and
    Engineering
Shri Mata Vaishno Devi University
Katra, India

Sahil Sholla
Department of Computer Science and
    Engineering
Islamic University of Science and
    Technology
Awantipora, India

Shafalika Vijayal
Department of Computer Science and
    Engineering
Model Institute of Engineering and
    Technology
Jammu, India

Rohit Wanchoo
Software Developer
TechAspect Solutions Pvt. Ltd.
India

Syed Rameem Zahra
Department of Computer Science and
    Engineering
National Institute of Technology
Srinagar, India

Shoaib Amin Banday
Department of Electronics &
    Communication Engineering
Islamic University of Science and
    Technology
Awantipora, India

1

Intrinsic Security and Privacy Vulnerabilities in the Internet of Things

Syed Rameem Zahra

CONTENTS

1.1 Introduction

The Internet of Things (IoT) has, over the last few years, significantly revolutionized many different facets of life.

IoT depends on combinations of different procedures; for example, identification, detection, networking, and calculation. It empowers large-scale developments in technology, and value-added services which customize clients' communication with different "things". Its spectrum of application spans from monitoring the dampness in crops to monitoring the movement of enemies in war zones. However, regardless of such colossal achievements, the near future does not guarantee an ideal abode for IoT (Zahra, S and Chishti, M, 2019a). This statement is based on the grounds that enough attention is not given to the security of IoT. Recent attacks are transforming the way companies view the S&P of IoT as even a tiny IoT device that has minimal capabilities presents significant hazards to the network's entire security infrastructure when its protection is violated. That is because by linking anything to the Internet, IoT provides a large attack surface for rogue players. They

1

can easily exploit the weak points and confidential data. It is therefore imperative that our outlook toward S&P of IoT changes to make it a necessity of the design process.

IoT depends on coordination of different protocols and enabling technologies. However, the differences between benchmarks, the fragmentation of standards, and the range of enabling technologies used, produce huge difficulties in giving full availability of everything. Security issues among the enabling technologies further complicate matters.

This chapter is structured as follows: Section 1.2 identifies the inherent features of IoT that mark its uniqueness, and differentiate it from other paradigms. It also explains how the features which bring about the success of IoT become a threat to its S&P. Section 1.3 describes major enabling technologies of IoT, and visualizes them from various security aspects like authorization, access control, trust, etc. Section 1.4 describes the protocols used to cater to the unique characteristics of IoT, and their S&P issues. Section 1.5 concludes the chapter.

1.2 Security Issues Arising from Inherent IoT Features

Features like low battery backup, constrained storage capacity, and minimal processing ability, define IoT devices (Zahra, S and Chishti, M, 2019b). As such, "constrained" comes as one of the inherent features of these devices. Apart from being constrained, the versatility of IoT devices comes from features like: interdependence, heterogeneity, plurality, and mobility (Wei Zhou et al., 2019). Table 1.1 describes these intrinsic features along with their security aspects. It also sketches out the research opportunities for researchers.

1.3 Enabling Technologies of IoT and Their S&P Issues

With the advent of IoT, sensors started to communicate over the Internet. To slash the cabling costs required for interconnecting billions of IoT devices, wireless communication must be preferred in sensors. Also, low power communication standards, emerging technologies, as well as service applications, need to develop harmoniously to match market demands to client needs. This section elaborates various enabling technologies that are of paramount importance to IoT, with special mention to their security aspects.

1.3.1 Radio Frequency Identification (RFID)

RFID empowers a panorama of applications in smart cities ranging from smart parking to healthcare applications, from asset management to tracking of objects (Gerhard, P et al., 2013). It consists of numerous tags attached to various objects and some readers which are accountable for communication with the tag and its powering (Gerhard, P et al., 2013). RFID makes identification possible from a distance, and contrary to the previously used barcode technology which requires line-of-sight operation; it does so without the need for line-of-sight.

The RFID tags can have any one of the three possible configurations: Active Reader Active Tag (ARAT), Active Reader Passive Tag (ARPT), and Passive Reader Active Tag (PRAT). Active tags need a power source, whereas passive tags are live only as long as the

TABLE 1.1

Inherent Features of IoT and Their Security Aspects

Inherent Feature	Description	Security Challenge	Research Opportunity	Probable Solution
Interdependence	IoT uses the IFTTT services to pilot automation of various applications. As such, when one thing happens, it leads to the occurrence of something else as well.	• Security related context is not taken into account. • Ambiguity by giving the feel that different applications and services operate independently.	• Analyze the behavior change caused by the cross-device IoT interdependence. • Design context based permission systems which would curb IoT devices from changing their behavior.	• The users must monitor their network and the permissions they give out. • Keep the firmware of the device updated at all times.
Heterogeneity	IoT uses varied type of firmware and interfaces. Also, there is a range in the kind of protocols used for access control, authentication and communication. This heterogeneity is advocated by the diversity of IoT functions.	• Vulnerabilities in protocols could be easily targeted. • More than 90% of IoT devices suffer from hard-coded key security challenges and more than 94% present web security vulnerabilities.	• Better IDS and IPS systems need to be built which take IoT heterogeneity into account. • Develop AC mechanisms which support dynamic nature of IoT environments. • Design a standard authentication and key management (AKM) technique so that pre-configured security information is not required and less pressure is put on constrained IoT devices.	• Use anomaly based IDS that work in real time, reporting anomalies. • Update firmware continuously.
Constrained	IoT devices face limitations in resources like battery, storage and computational ability. Also, in some applications, their function is to work under deadlines, thereby constraining them by time as well. • Usually used microcontrollers of IoT devices have less than 2MB RAM.	• IoT devices cannot run necessary operations like memory isolation and ASLR for the lack of a MMU. • They cannot run existing heavy weight encryption and authentication algorithm. • A simple anti-virus like common touch needs 128MB RAM.	Design light weight encryption and authentication schemes.	Use multi-layer defense architecture to make carrying out of the attack cumbersome.

(Continued)

TABLE 1.1 (Continued)

Inherent Features of IoT and Their Security Aspects

Inherent Feature	Description	Security Challenge	Research Opportunity	Probable Solution
Plurality	There is an estimation of 50 billion IoT devices by 2020, meaning IoT devices will be everywhere.	• Botnet attacks could be launched because of this huge attack surface. Examples of recent attacks include names like Mirai, Reaper, Hajime and Persirai etc. • Scalability issues.	Improvisation and enhancement of Internet infrastructure for IoT services.	• User must know the network and periodically run in-depth forensic analysis over the organization. • Flag outgoing suspicious traffic. Incoming traffic is usually kept under check. Same should be followed for outgoing traffic because any ransomware cannot get off the ground without connecting with its CandC which lies outside. • Update firmware.
Mobility	Any device can be made into an IoT device which implies the majority of them could be mobile.	• For establishing connection to Internet, the mobile devices usually employ less secure wireless connections leading to enhanced error rates and reduced bandwidths. • Since they can join various networks attackers are tempted to load them with malicious software for rapidly propagating the malicious code.	Design mobility resilient security protocols for IoT.	Know your network and run its in-depth forensic analysis.

energy from a reader source is applied. The implied advantage of passive tags over active tags is that they require neither batteries for operation nor maintenance. Active tags, on the other hand, need both. In PRAT, the reader is passive while the tag is active; therefore reader obtains energy from the battery of active tag. In ARAT, both the reader and tag are active where the tags wake up only when they come in the reader's proximity. In ARPT, reader is active and the tag is passive.

A passive tag has three main components: an antenna, a semiconductor chip connected to antenna, and a type of encapsulation. The antenna captures the information and relays it to the chip. The third component—encapsulation—maintains the integrity of the tag by keeping the antenna and chip in correct form, protecting them from environmental conditions. For transferring power from reader to tag, there exist two distinct approaches: electromagnetic wave capture, and magnetic induction. Depending on the type of the tag, both designs can send 10μW–1mW to sustain tag operation. The passive tags are very cheap, making them a best choice for deployment of ubiquitous sensing (Gerhard, P et al., 2013).

Security aspect: One of the most common information types that is stored on the tag is Electronic Product Code (EPC), which helps in the unique identification of tags. EPC marks is one of the inherent security weaknesses of the RFID technology, since we rely on EPC to tell us if the product is genuine or fake. For example, an attacker outfitted with a good reader can examine numerous RFID labels to collect countless EPC numbers. S/he at that point can generate RFID labels which give out precisely the same EPCs gathered. These tags are called "cloned tags". The cloned tags can be connected to fake things which ought to be perceived as genuine. In addition to cloned tags, RFID raises numerous other risks of breaches to security and safety. These are identified as:

- Attack on access control: The RFID technology doesn't have any authentication procedure (Ivan, C et al., 2018). This lack of authentication gives rise to the risk of unauthorized disclosure of data.

- Attack on integrity: Once the attacker gains access into the data, s/he can easily modify the content.

- IoT environments that make use of RFID technology are also prone to attacks like eavesdropping, Denial of Service (DoS), man-in-the-middle, spoofing attacks, and others[1] (Borgohain T et al., 2015).

- Attack on privacy: Since the EPC numbers are unique; the customers could be tracked down if s/he carries any product whose RFID tag has not been removed. For example, a customer buys a clothing item from a store. The salesman with a criminal mindset skips to remove the RFID tag. The customer can then be tracked every time s/he wears the item; the seller can know exactly his/her shopping pattern, and the purchasing power. As such, the customer can become a potential target for theft (Roy W, 2006).

1.3.2 Near Field Communication (NFC)

The easiest approach to implement a passive RFID system is to make the use of NFC, the basis of which is formed by Faraday's Principle of Electromagnetic Induction. It is a short range, low power, wireless connection that is able to transmit small amounts of data between two devices. Unlike Bluetooth technology, initial pairing is not needed for sending data, therefore it is fast. The operational frequency band of NFC is 13.56 MHz and contrary to the RFID technology, NFC offers bidirectional communication (Gerhard, P et al., 2013). The integration of NFC to smart phones allows it to be used in smart cities as well (Saber, T et al., 2017). In fact, it is considered to be one of the key enabling technologies of smart cities. A worthy to mention NFC application is the digital wallet where a smart phone with NFC can be used as a replacement for various cards such as bank cards, public transportation cards, identification cards, etc. Some other real world use cases of NFC include: occupancy detection in homes, smart energy metering[2], collection and control of data (Opperman and Hancke, 2012), touristic surfing of a city by navigating scattered smart posters, and smart parking[3]. Ultimately, NFC technology will integrate all these services into one single mobile phone. However, NFC suffers from vulnerabilities which threaten user privacy and system security.

Security aspect: The NFC technology related attacks can be classified as those on the entire NFC systems or just the NFC components as shown in Figure 1.1. These security issues are explained in more detail in the following sub-sections.

- *Security issues in NFC systems*: Among the most important assets that require protection in NFC based systems is the user data privacy. The data that resides on the NFC's host controller (Madlmayr, G. et al., 2008). Also, if the host controller is compromised, the entire operability of the NFC can be taken over, and the data stored on the NFC tags can be changed. Moreover, the communication in NFC occurs over the RF links which suffer from inherent security weaknesses. This gives rise to threats of attack on communicated information, i.e., attacks on data integrity (Madlmayr, G. et al., 2008).

- *Security issues with data on NFC tags:* The tags are components of pivotal importance for applications operating in reader/writer modes. Several attacks that can be carried out on the NFC tags have been reported in the literature: those that try to manipulate the tag data; others which attempt to replace a genuine tag by a malevolent one, assaults which crack the tag's write protection and overwrite its contents, and cloning of NFC tags (Mulliner, C., 2009). Rieback, M et al. (2006) say that since NFC systems are based on RFID systems, RFID-based SQL injection attacks as well as RFID worm and virus attacks are conceivable on NFC tags. Moreover, the DoS attacks could be implemented on NFC tags by manipulating the NFC Data Exchange Format (NDEF) message and putting it on the top of NFC-tag which is associated with a

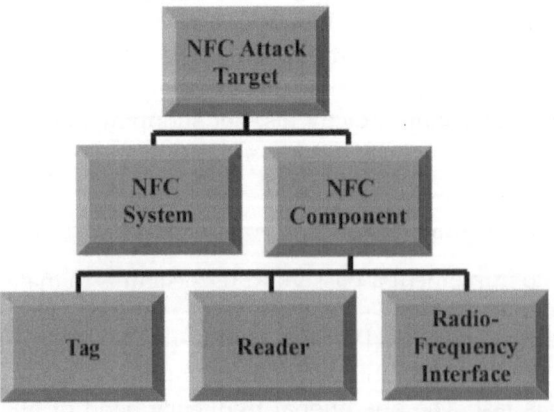

FIGURE 1.1
NFC Components.

service that the attacker wishes to bring down. An example of an NFC DoS attack would be if every time the customer touches the NFC-tag, his/her mobile phone crashes and reboots. When such a mishap occurs, a non-technical user is bound to think that the NFC-tag is crashing the phone, and subsequently stops using the service (Mulliner, C., 2009).

- *Security issues with NFC readers:* Reader is a vital component of NFC devices as it facilitates the applications that require card emulation mode. However, these NFC readers suffer from the danger of theft, destruction and removal. Moreover, they hold sensitive information like encryption keys. As such, the NFC readers could be a huge temptation for the attackers; given the type of function they have and the kind of information they hold. Therefore, it is extremely important to explore the susceptibilities of NFC readers and propose solutions to deal with them. Unfortunately, the security of NFC readers hasn't been subjected to sufficient research yet.

- *Security issues with RF interface:* NFC employs interfaces based on wireless communication. Wherever the wireless communication is used, there is a possibility of attack (eavesdropping, man-in-the-middle, DoS, relay, phishing, data corruption, modification, insertion attacks). NFC is no exception.

- *Lack of uniform encoding standard:* Japan supports Universal Identification (UID) whereas Europe follows the EPC standard. As such, no uniform encoding standard is being followed for the identification of items. Consequently, readers may not always be able to gain access easily to the tag data, and errors may occur in the process.

- *Data collision:* It is quite possible that multiple tags send data to the reader simultaneously, leading to incorrect interpretation of data by the reader.

- *Trust:* Proper trust must exist among the readers and tags, and between the readers and base stations.

1.3.2.1 Wireless Sensor Networks (WSN)

WSNs showcase a dynamic environment consisting of self-organizing and multi-hop networks. They are typically constrained of storage, computation and sensing range abilities. These constraints give rise to numerous security issues. From the IoT perspective, the perception layer is meant for data collection, the basis of which if formed by WSN, offering a potentially huge number of sensing nodes. However, these WSN sensors have limited scope which makes the network structure complex. Moreover, in the process of data collection, data may be subjected to various security threats like tampering, routing attack, eavesdropping, etc. Data security issues may be grouped into four broad categories:

1. *Confidentiality of data:* To prevent access by unauthorized parties to the critical information.
2. *Authenticity of data:* Establishing that a message is exactly what it claims to be.
3. *Data integrity:* To ensure that data has not been modified, is consistent, and complete.
4. *Data freshness:* To ensure that old and stale messages are not replayed. Instead, the copy of the data is most recent.

TABLE 1.2

Issues in WSN Security Solutions

Category	Security Issue	Inference
Asymmetric cryptography (Ahmad, M, 2018)	Storage and computational ability constraints do not encourage the use of this technique in WSNs.	Resource intensive cryptographic technique.
Symmetric cryptography	Fewer simple calculations required.	A good option for WSN.
Symmetric cryptography key exchange protocol (Ahmad, M, 2018)	The protocol is extremely complex which makes the scalability of this cryptographic technique over the WSNs quiet small.	Either a lightweight key exchange protocol needs to be developed, or a compromise between symmetric and asymmetric techniques needs to be achieved.
Confidentiality of key in symmetric cryptography	WSNs are often deployed in environments that remain unattended for long durations.	The entire WSN network comes under a security threat when even a single node is compromised.
Authentication mechanism in symmetric cryptography	Authentication in WSN is achieved by Message Authentication Code (MAC), which is a computation-heavy procedure.	This leads to heavy storage space, communication and power consumption needs.
Key distribution	Secure transportation and distribution of the key to legitimate users.	Lightweight key distribution mechanism needs to be developed to support resource constrained devices.
Routing protocol	Wormhole, sinkhole, and DoS attacks are possible.	The routing attacks bring the entire network down. There is a need for development of secure routing protocols.
Node trust	Management of trust update poses security threats in WSNs.	Efficient trust management mechanisms are needed to secure the WSNs.

These issues are respectively managed by encryption (hiding the plain text), key management (includes generation of a secret key, its distribution, storage, updating and destruction), secure routing, and node trust. Table 1.2 demonstrates the problems associated with each of these solutions in WSNs.

1.3.3 Machine to Machine Communication (M2M)

M2M represents one of the key technologies for building IoT networks. It involves the communication that takes place without any human intervention between computers, smart sensors and actuators, processors, and the various other mobile devices, in an autonomous manner (Esfahani, A et al., 2019). This autonomous behavior is of paramount importance to pilot the working of various IoT applications like home automation, healthcare, manufacturing, etc. Industrial IoT (IIoT) particularly represents one of the most striking applications of M2M communications, where it allows man, machine and product to remain connected throughout the process of production up to the managerial level (Esfahani, A et al., 2019). As such, IIoT and M2M together are at the brim of bringing the Industry 4.0 revolution (Colombo, A, et al., 2017). M2M has four main components: sensing, heterogeneous access, processing of information and application and processing.

Security aspect: Security becomes a challenge in M2M because of its diverse deployments. Attacks like DoS could be easily launched on these deployments, and thus there would be a significant compromise on their integrity, authorization, authentication, availability, and confidentiality (Esfahani, A et al., 2019). It is extremely important to provide end-to-end security here, as otherwise the result could be another "Mirai botnet"-like attack of 2016 that used more than 600,000 less-secure devices to launch a large-scale Distributed Denial of Service (DDoS) attack, generating an attack traffic rate of 1 Tbps to bring down various servers. The effect of such an attack could be quantified by the $110 million of lost revenue.

1.3.4 Vehicle to Vehicle Communication (V2V)

This occurs between vehicles that form the nodes in the vehicular ad hoc network environment. The vehicles communicate with each other using various protocols to share information related to driver and passenger safety, road safety, passenger comfort, etc. The concept of "Smart Transport", which is recognized as one of the primal applications of IoT by the European Research Cluster on the Internet of Things (IERC), is unachievable without the understanding of V2V communications (Zahra, SR and Chishti, MA, 2019a). When a smart car makes use of V2V communication in a situation of danger, it sends useful warnings to other vehicles, recommending them not to approach that region (Zahra, S, 2018). This type of communication can be single-hop or multi-hop. Safety warnings are sent using the rapid single-hop V2V approach while as the non-safety messages are broadcasted using the multi-hop V2V communication.

Security aspect: The issue of security of V2V communications is highlighted in Table 1.3 (Zahra, S, 2018).

1.3.5 Low Rate Wireless Personal Area Network

It is a short-range radio technology that is best suited for low-power and lossy networks owing to the advantages that it provides including: significantly reduced power consumption, battery lifetime improved to as much as ten years, and range that spans up to 15 km. As per the 802.15.4 standard, Low Rate Wireless Personal Area Network (LR-WPAN) offers inexpensive and low-rate communication for sensor networks, and besides having the upper layer protocols like 6LoWPAN and Zigbee, it has physical and medium access layer protocols as well. As a result of all its advantages, LR-WPAN is one of the foundational technologies of IoT.

Security aspect: There is an inherent threat of interference and interception among wireless communications. Moreover, the devices working in LR-WPANs are resource constrained, making the working of security services a key issue. Attacks that are based on resource consumption could be easily launched on these devices (Jianliang, Z, et al., 2006). Table 1.4 summarizes various attacks that are possible in LR-WPAN environments.

TABLE 1.3

Security Issues in V2V Communication

Attack	Attack on	Description
DoS	Availability of server	• Attackers send huge numbers of control messages to vehicle onboard units (OBUs). • Because of memory constraints, they cannot handle this pressure, and the VANET is effectively brought down.
Black hole	Availability of a service/server	• The attacker node lures other vehicles to pass it the messages, as it has the shortest link to the roadside-unit (that connects the VANET to the Internet). • After receiving the traffic, it drops all of the packets. • If messages sent to the attacker vehicle were safety warnings, and they were dropped, the consequences could be disastrous.
Jamming	Availability of communication channel	• The communication channel is disrupted. • Signal–noise ratio lowered at the receiving end.
Greedy behavior	Services provided by MAC layer	• The attacker node punishes the honest ones by trying to get a faster access into the medium, ignoring every medium access protocol.
Sybil	Authenticity	• The corrupt node takes up multiple fake identities. • Particularly dangerous in VANET environments as the entire system can be confused.
Eavesdropping	Confidentiality	• Passive attack. • Silently listens to entire conversation, which might be necessary for vehicle tracking operations.
Spoofing	Authenticity, Privacy	• Attackers send illegitimate messages with incorrect details to disrupt vehicular communications
Man-in-the-middle	Confidentiality, Integrity, Authenticity	• Network traffic is directed to an attacker node in order to manipulate the information
Masquerading	Integrity	• The attacker takes a false identity (such as an authentic user). • Sends false messages into the system, pretending to be someone else.
Loss of event track-ability	Non-repudiation/Accountability	• After sending or receiving the message, the attacker denies doing so.

TABLE 1.4

Attacks in LR-WPAN

Attack	Attack Layer	Description	Impact in LR-WPANs
Jamming	Physical	In this attack the communication channel is jammed by emitting high-power RF signals without following any standard protocol; similar to DoS attacks.	The extremely less transmission power of LR-WPANs makes them highly vulnerable to jamming attacks. Moreover, particular LR-WPANs normally operate on one frequency channel instead of exploiting the available 27 channels (Jianliang, Z, et al., 2006).

(Continued)

TABLE 1.4 (Continued)

Attacks in LR-WPAN

Attack	Attack Layer	Description	Impact in LR-WPANs
Hostage	Hardware	The device is hacked and tampered with. Hostage attacks are difficult to avoid in LR-WPANs (Jianliang, Z, et al., 2006).	If extensive security is implemented on each and every device, the primary design goal of low cost will be defeated.
Exhaustion	Physical and MAC sublayer	Type I: The attacker node tries to link with every co-coordinator present in its reach, breaching the protocol specifications (Jianliang, Z, et al., 2006). Type II: Hack the coordinator and use it to lure multiple surrounding nodes to use it as their coordinator by offering better quality links and low levels in the tree.	Force the devices to stay awake by transmitting maliciously configured beacons. This results in devices using all their battery power.
Collision	MAC sublayer	Type I: Collision with acknowledgment frames. Such collisions would require the source node to unnecessarily move into the exponential-back-off phase, leading to delays. Type II: Collision with association response frames. Such collisions force the device to start the multi-step association process right from the start (Jianliang, Z, et al., 2006). Type III: Collision with beacon frames. This type of collision will cause the children of various such nodes to get orphaned.	Collision attacks are extremely tough to detect in LR-WPANs when the attacker becomes selective; the attack seems to be a ransom collision rather than an intentional one.
Route disturbance by a corrupted node in LR-WPAN cluster	Network	The attacker continuously sends association requests to the coordinator every time with a new fake device address (Jianliang, Z, et al., 2006).As such, the coordinator reaches its full capacity, after which it rejects every genuine request.	This attack is dangerous when it is launched near the root of the tree.
Route disturbance by a corrupted coordinator in LR-WPAN cluster	Network	A corrupted coordinator can bring down the devices by keeping them active all the time.	Dropping and modification of packets will happen. The devices will be drained of energy.
Looping in LR-WPAN cluster in trees	Network	A corrupt coordinator can try to form loops in acyclic trees of LR-WPAN clusters resulting in routing loop problem (Jianliang, Z, et al., 2006).	The packets would loop among the children of the nodes. Such attacks would lead to resource consumption attacks.

1.4 Infrastructure Protocols of IoT and Their S&P Issues

The links utilized for communication in IoT have a heavy loss rate, low throughput, and lack node co-operation. In addition, the traffic patterns are not always construed by a point-to-point schema. They can both be point-to-multipoint or multipoint-to-point patterns, and hence the prevailing protocols cannot deal with these needs. Consequently, an entire stack of standardized protocols was developed including the IEEE 802.15.4 standard protocol for Wireless Personal Area Networks (WPAN) and the IPv6 over Low Power Personal Area Networks (6LoWPAN) protocol for providing IPv6 connectivity to these networks. At the routing layer, to cater to the unique characteristics of these networks, Internet Engineering Task Force (IETF) specified a standard routing protocol called Routing Protocol for Low-power and Lossy Networks (RPL) based on IPv6. This section defines the most important infrastructure protocols along with their S&P issues.

1.4.1 Routing Protocol for Low-power and Lossy Networks (RPL)

It is an IPv6-based distance-vector de facto IoT routing protocol (Winter, T et al., 2012). It gives advantages like minimal overhead, energy efficiency, and optimal routing, making it the best choice for constrained networks. RPL works by finding routes the moment it comes into service. This feature classifies it in the category of proactive routing protocols. The devices using RPL are organized using Destination-Oriented Directed Acyclic Graph (DODAG) that is a blend of mesh and tree topologies (Winter, T et al., 2012). The IETF's RPL specification lacks the proposal of any proper and significant security models. In essence, the existing RPL standard employs key management scheme for authenticating the devices entering into the RPL network.

Security aspect: The current RPL standard lacks a model defining how to achieve the authentication and secure network connection among network devices working on various security critical missions. This factor makes the devices prone to a number of routing attacks (Tsao, T, et al., 2015) (e.g., routing table falsification attack, sinkhole, wormhole, selective forwarding, Sybil, worst parent attack, routing information replay attack, rank attacks, DAO-inconsistency attack, and other Byzantine attacks). Also, the constrained nature of RPL networks makes them vulnerable to these attacks. The encryption based solutions won't work in the situations where the authorized internal nodes become corrupt and launch the attacks. Also, some security modes of RPL essentially require public key cryptography, which is too expensive to use regularly.

1.4.2 6LoWPAN

An IETF standardized protocol RFC 6282, 6LoWPAN is designed to provide IPv6 connectivity for heavily constrained devices over low-power and lossy networks, and as such has considerably revolutionized the IoT landscape by pursuing the expansion of IPv6 services to smart and puny devices (Glissa, G, and Meddeb, A, 2019). Since it provides the IP connectivity over low-power and lossy networks, it is viewed as a foundation stone for IoT network build up. The various IoT use cases like smart cities, smart homes, smart offices, etc., utilize 6LoWPAN for IP communication as it employs a frame size of 127 bytes-one that is optimal for the low power sensor devices used in these cases. For larger IPv6 packets, 6LoWPAN uses the data link layer protocol IEEE 802.15.4. Additional support to fragmentation is provided by 6LoWPAN at the adaptation layer, implying these fragments require some sort of buffering and processing before they can be forwarded. All the processes meant for carrying out the fragmentation (e.g., buffering, processing, and forwarding of fragmented packets) become resource-expensive for the already constrained IoT devices. A gateway device popularly called 6BR (6LoWPAN Border Router) is responsible for connecting the 6LoWPAN network to the Internet, and executes the processes of fragmentation/assembly and compression/decompression of 6LoWPAN packets to forward them between the Internet and 6LoWPAN network.

Security aspect: Being a successor of Internet protocols IPv4 and IPv6, 6LoWPAN easily inherits their security threats. Moreover, limitation on resources, absence of proper standards, low experience of professionals on 6LoWPAN implementation, and inadequate administrative proficiency make 6LoWPANs susceptible to all kinds of old and new network attacks (Monali, M, and Krishna, A, 2017). Moreover, the customary security checks and solutions are profoundly inadequate for 6LoWPANs', requiring design of new countermeasures to ensure refuge against such easily understood network assaults. It utilizes RPL for routing purposes by employing the control message "Neighbor Solicitation" (Monali, M, and Krishna, A, 2017) and 6LoWPAN-Neighbor Discovery (6LoWPAN-ND) for the purpose of neighbor discovery by using the control message "Neighbor Advertisement" (NA). If the attacker is a trusted insider, it can exploit these control messages easily to launch different types of attacks (e.g.. the IPv6 Spoofing attack).

1.4.3 IEEE 802.15.4

IEEE 802.15.4 is the de facto protocol used for the communication in WPANs. Because of the small Maximum Transmission Unit (MTU) of 127 bytes, the low-power and lossy networks based on 6LoWPAN find the incorporation of its adaptation layer above the link layer of 802.15.4 to be the best fit (Minhaj, M, and Khaled, S, 2018). This will endow the low power and low-rate sensor nodes with high IP based communication competence.

Security aspect: Given the vulnerabilities of 6LoWPAN to various attacks, security is one of the most critical aspects that require attention in 6LoWPAN based IoT networks. This security for the communication among the internal nodes of the network is provided by hinging the protocols that adopt 6LoWPAN on the security sublayer of 802.15.4. The security sublayer attains this goal by stopping any internal malicious node inserting its frames into the network by appending a frame control and a Message Integrity Code (MIC) to every frame, thereby ensuring essential security requirements such as integrity, authentication, newness of frames; and to certain extent, confidentiality as well (Xiao, Y, et al., 2006). The auxiliary security header field provides information specific to security and has three sub-fields to it: Security control, Frame control and Key identifier is shown in Figure 1.2.

The security control field maintains the information about the kind of security that needs to be applied on the frame. The purpose of the key ID in the design remains uncertain. Every

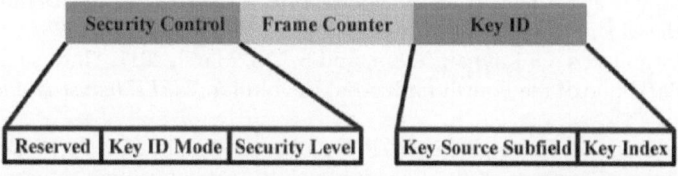

FIGURE 1.2
Auxiliary security header format.

node in the network is pre-loaded with the shared key. This makes these nodes highly vulnerable to a variety of attacks. All the attacker needs to do is to physically intrude into the node to extract its cryptographic details. For dealing with this type of attack, tamper-proof hardware can be used, but it becomes expensive and does not guarantee unlimited security. If the attacker is successful in compromising the node, it can easily load the infectious frames onto it as well as add other non-authorized devices into the network of the victim device. Moreover, as the upper layer protocols depend on 802.15.4 for its security sublayer, the error and security loophole can travel to those layers as well (Winter, T, et al., 2012).

1.5 Conclusion

Excessive advances in the miniaturization of devices and communication technologies led to the creation of the IoT paradigm. However, the same elements which fed into the success of IoT brought along with them a handful of S&P issues. This chapter covered S&P vulnerabilities which come into the picture of IoT because of the intrinsic features of IoT that make it special, the technologies which enabled its creation, and the protocols on which it runs. It also highlights the major research opportunities which could be taken up, and the need for security professionals to up their game with IoT to enable it to become more secure, and better established.

Notes

1 These attacks are discussed in Table 1.3.
2 In which meters are read by NFC enabled smart phones and the bills are paid automatically by sending the readings to the banking backend.
3 In which NFC enabled phones can be utilized to act as e-tickets for making an entry into the parking as well as an e-wallet for making payments for it.

References

Ahmad, M., "Alleviating Malicious Insider Attacks in MANET using a Multipath On-demand Security Mechanism," 2018, *I.J. Computer Network and Information Security, MECS Press,* 6, 40–51.

Borgohain, T., U. Kumar, and S. Sanyal, "Survey of Security and Privacy Issues of Internet of Things," 2015, *International Journal of Advanced Network Applications,* 6(4), 2372–2378.

Colombo, A., S. Karnouskos, O. Kaynak, Y. Shi, and S. Yin, March 2017, "Industrial Cyberphysical Systems: A Backbone of the Fourth Iindustrial Revolution," *IEEE Industrial Electronic Maazine,* 11(1), 6–16.

Esfahani, A., M. Georgios, M. Rainer, B. Firooz, R. Jonathan, B. Ani, M. Silia, G. Markus, S. Christoph, and B. Joaquim, February 2019, "A Lightweight Authentication Mechanism for M2M Communications in Industrial IoT Environment," *IEEE Internet of Things Journal,* 6(1), 288–296.

Gerhard, P. Hancke et al., 2013, "The Role of Advanced Sensing in Smart Cities, *Sensors*, 13(1), 393–425.

Glissa, G. and A. Meddeb, 2019. "6LowPSec: An End-to-End Security Protocol for 6LoWPAN," *d Hoc Networks, Elsevier*, 82, 100–112.

Ivan, C., Miroslav Vujić, and Siniša Husnjak, 2018. "Classification of Security Risks in the IoT Environment," *Far East Journal of Electronics and Communications*, 18(6), 915–944.

Jianliang, Z., Myung J. Lee, and A. Michael, October 2006, Toward Secure Low Rate Wireless Personal Area Networks," *IEEE Transactions on Mobile Computing*, 5(10), 1361–1373.

Madlmayr, G. et al., 2008, "NFC Devices: Security and Privacy," *Proceedings of Third International Conference on Availability, Reliability and Security*, Barcelona, 642–647. https://doi.org/10.1109/ARES.2008.105

Minhaj, A. and S. Khaled, 2018, "IoT Security: Review, Blockchain Solutions, and Open Challenges," *Future Generation Computer Systems*, 82, 395–411.

Monali, M. and A. Krishna, 2017, "Modeling and Analyses of IP Spoofing Attack in 6LoWPAN Network," *Computers & Security*, 70, 95–110.

Mulliner, C., 2009, "Vulnerability Analysis and Attacks on NFC-enabled Mobile Phones," *International Conference on Availability, Reliability and Security*, Fukuoka, Japan. https://doi.org/10.1109/ARES.2009.46.

Opperman, H., May 2012, "Smartphones as a Platform for Advanced Measurement and Processing," *Proceedings of 2012 IEEE Instrumentation and Measurement Technology Conference (I2MTC)*, Graz, Austria, 13–16. https://doi.org/10.1109/I2MTC.2012.6229588

Rieback, M., B. Crispo, and A. S. Tanenbaum, 2006, "Is Your Cat Infected with a Computer Virus?," *Fourth IEEE International Conference on Pervasive Computing and Communications PerCom*, 169–179. https://doi.org/10.1109/PERCOM.2006.32

Roy, W., 2006, "An Introduction to RFID Technology,"*IEEE CS and IEEE ComSoc, Pervasive Computing*, 5(1), 25–33.

Saber, T. et al., 2017, "A Review of Smart Cities Based on the Internet of Things Concept," *Energies*, 10(4), 1–23.

Tsao, T., R. Alexander, M. Dohler, V. Daza, A. Lozano, and M. Richardson, 2015, "A Security Threat Analysis for the Routing Protocol for Low-Power and Lossy Networks (RPLs)," Internet Eng. Task Force (IETF).

Winter, T., P. Thubert, A. Brandt, J. Hui, R. Kelsey, P. Levis, K. Pister, R. Struik, J. Vasseur, and R. Alexander, 2012, "RPL: IPv6 Routing Protocol for Low-Power and Lossy Networks," Internet Eng. Task Force.

Xiao, Y., H.-H. Chen, B. Sun, R. Wang, and S. Sethi, 2006. "Mac Security and Security Overhead Analysis in the IEEE 802.15.4 Wireless Sensor Networks," *EURASIP Journal on Wireless Communications and Networking*, 2006, 1–12.

Zahra, S., 2018, "MNP: Malicious Node Prevention in Vehicular Ad hoc Networks," *IJCNA*, 5(2), 9–21.

Zahra, S. and M. Chishti, 2019a, "Assessing the Services, Security Threats, Challenges and Solutions in the Internet of Things," *Scalable Computing: Practice and Experience*, 20(3), 457–484.

Zahra, S. R. and M. A. Chishti, 2019b, "Ransomware and Internet of Things: A New Security Nightmare," *IEEE 9th International Conference on Cloud Computing, Data Science, and Engineering (Confluence 2019)*, Noida, India. https://doi.org/10.1109/CONFLUENCE.2019.8776926

Zhou, W. et al., 2019, "The Effect of IoT New Features on Security and Privacy: New Threats, Existing Solutions, and Challenges Yet to Be Solved," *IEEE Internet of Things Journal*, 6(2), 1606–1616.

2

Loopholes in IoT Security Services

Shafalika Vijayal and Salim Qureshi

CONTENTS

2.1 Introduction

Physical security is the assurance of life and property and incorporates things as assorted as individuals, equipment, programs and even the information that happens from an occasion that causes loss.

A break in the physical security course of action can be possible with basically no knowledge by the aggressor. Regardless, the IoT and interconnectivity is profoundly influencing the physical security of industry. This carries two queries to mind: how to associate physical security gadgets to the Internet and guarantee they are distant from programmers that can hack the information, and how to utilize the present techniques that control and interrupt the gadgets existing as of now. Performing successive data and information backups and chronicling keeps duplicates of the information, and shields them from actual harm. Verifying the backup of information (data) ought to be completed by actualizing suitable security systems, allotting access rights to only the approved people, storing the backups off-site, having control of physical access to that storage, and utilizing encryption methods. Performing successive data and information backups keeps the duplicates safe from hacking, and covers them from both physical and technical harm. Protecting the physical device and the entry to their interfaces ought to be considered to carefully screen the vulnerabilities while putting away the information of a number of gadgets associated over the Internet and Cloud. IoT environments are vulnerable to attacks, with security gaps in key areas such as:

- Obsolete operating systems.
- Unencrypted passwords.
- Remotely accessible devices.
- Invisible indicators of threats.
- Direct Internet connections.
- No casual AV updates.

This does not mean that nothing can be done; consider the earlier statistics from CyberX which has studied over 3000 networks of IoT. The risk of exposure of personal information through implementation of smart homes, smart cars, smart healthcare, etc. increases by the expansion of IoT domain. Some IoT security issues include:

- Abla El Bekkali et al.[2] state that in 2014, Stuxnet attack affected the Iranian Nuclear Project.
- In 2016, the Jeep SUV hack was listed listed by Abla El Bekkali et al. [2] state [2].
- In 2016, a malevolent botnet was featured by the security specialists of Sucuri stated by Abla El Bekkali et al. [2].

As listed above, the security challenges for IoT are extreme. It is necessary to define valid and robust security architecture by keeping in mind that the regulations and policies related to IoT will keep changing.

In the chapter 'Loopholes in IoT Security Services' we will discuss the various vulnerabilities in physical devices connected over the IoT network which pose a threat to personal information, as well as the hardware through which transmission over a network occurs, and also list some countermeasures to deal with the various vulnerabilities penetrated by malevolent attackers.

2.2 Insecure Web Interface

A web interface is the platform of interaction provided between the user and the application that is running on the Web server. It may also be rendered as a programming connection to the Web. A Web interface is the most common environment in which a user can access any popular application or webpage through any system software (Windows, Linux, Macintosh, Oracle Solaris etc.) underneath. The interface provided by the Web can be any connection that makes a user has its social presence, e-commerce trading, e-education, e-tenders. For example, your smartphone provides a web interface through various applications installed on it, for which all you need is a username and password to access and control settings. To deal with IoT, an actuator and setting panel is required. Most of the IoT devices already have a web interface established with the server. As Web interfaces have been in use a long time, security is the prime concern while implementing them. IT professionals have long been struggling with the methodologies to implement secure interfaces and improving them in the long run.

Let us consider a home security system. You leave your home at the beginning of the day. At your office, you recollect that you didn't turn the alert framework on, so you sign in to the web interface and turn it on; all thanks to the Internet. Be that as it may, on the off chance that it very well may be diverted on from remote, who's to state nobody else can turn it off, and even open the doors or windows?

How to improve the situation:

- Never allow the default passwords to be used: A password change option on setting up of the device is mandatory.
- Prevent brute-force attacks: Always include an access limit and freeze the account after a specified number of failed accesses. However, include a reset password feature for non-Web interaction with the device.
- Ensure that the program is not prone to vulnerabilities such as XSS, CSRF, and SQLi etc.
- Make sure not to share the credentials, and do not expose them over the network.
- Always keep the security updated to latest encryption algorithms and techniques to avoid any vulnerability as stated by Mookyu Park et al. [4] in their work.

Specific security vulnerabilities that could lead to unauthorized privileges to the malicious user include: account enumeration, weak default credentials, cross-site scripting, SQL injection, session management, and weak account lockout settings.

2.2.1 Account Enumeration

Account enumeration happens when an evil user applies a brute-force algorithm to make guesses to gain access to the system by proving to be an actual user in the system. Account enumeration is one such type of web application vulnerability listed by Mohit Mittal et al [5]. Two easily targeted fields through which account enumeration takes

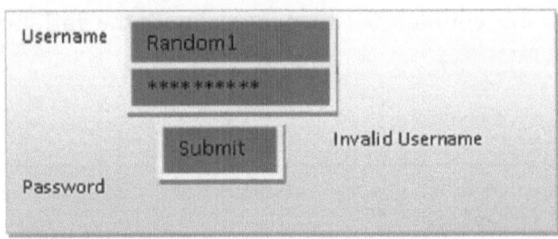

FIGURE 2.1
Account enumeration.

place are in a "Login" page and "Forgotten Password" section. The malicious actor seeks for variations in the server's reply depending on the validity of details entered. The login form is the easily targeted spot for such attacks. When the user enters his/her username and password, and the username entered is invalid, the server responds by the statement that the entered username doesn't exist. A malevolent actor would understand easily that the problem exists with the username, but that the password entered is correct. Figure 2.1 illustrates account enumeration through invalid login details.

2.2.2 Weak Default Credentials

An attacker could gain and use the following if the credentials are accessible to him/her:

- Approval to view the information present on the resources protected by the passwords.
- The admin panel rights such as "dashboard" or "management console", enabling complete access to the application.

2.2.3 Cross-site Scripting (XSS)

Cross-site scripting (XSS) is a customer-side code injection attack. The aggressor attempts to float harmful content in an Internet browser of the unfortunate casualty by reminding them of the malignant code for a genuine site page or web application. The actual assault happens through the unfortunate casualty visiting the site page or web application that runs the vindictive program. The web page or application becomes a vehicle to convey the harmful content to the client's program. Seemingly powerless interfaces that are normally used for cross-website scripting assaults such as discussions, message sheets, and site pages that permit remarks.

A site page or web application is powerless toward XSS on the off chance that it uses unfiltered client contribution to the yield that it creates. XSS assaults are conceivable in VBScript, ActiveX, Flash, and even CSS. In any case, they are necessary in JavaScript, in light of the fact that JavaScript is central to most perusing encounters.

Is the cross-site scripting the user's problem? In the event that an aggressor can manhandle XSS defenselessness on a webpage to execute self-assertive JavaScript in a client's program, the security of that website or application and its clients has been undermined. XSS isn't the client's concern like some other security breaches. On the off chance that it is influencing your clients, it is influencing you.

Cross-webpage scripting may likewise be utilized to damage a site as opposed to focusing on the client. The assailant can utilize infused contents to change the substance of the site or even divert the program to another site page, for instance, one that contains harmful code.

2.2.4 SQL Injection

SQL injection (SQLi) is a sort of an infusion attack that makes it possible to execute malignant SQL clarifications. Attackers can use SQL injection vulnerabilities to avoid application security endeavors. They can claim approval and endorsement of a webpage page or web application and recoup the substance of the entire SQL database. They can use SQL injection to incorporate, modify, and eradicate records in the database. A lack of protection against SQL injection may impact any webpage or application that uses a SQL database; for instance, MySQL, Oracle, SQL Server, etc. Criminals may use it to increment unapproved access to your delicate data: customer information, singular data, trade transactions, ensured advancement, to say the very least.

2.2.5 Session Management

Session management is the ruleset that sets interactions between web applications and users. Programs and sites use HTTP to convey, and a web session is a progression of HTTP solicitations and reaction exchanges made by a similar client. Since HTTP is a stateless convention, where each solicitation and reaction pair is autonomous of other web co-operations, each direction runs freely without knowing past courses. To present the validation and access control (or approval) modules ordinarily accessible in web applications.

There are two sorts of session management: cookie-based and URL rewriting. These can be utilized freely or together. A web overseer uses session management to follow the patterns of repeated visits to a site and changes inside the website.

2.2.6 Weak Account Lockout Settings

Weak account lockout settings are applied to all the applications which perform authentication or use a third-party authentication module. When the malicious attackers try to gain access to the user account by attempting a large number of possible passwords, lockout policy of authentication is used. This kind of attack leads to denial of service, due to which the owner of the account may be denied access to his/her account from exhaustion of allowed login attempts. This attack is especially powerful if the attacker can guess a large number of usernames or account numbers for a scripted attack on many users.

Following are some of the improvements against the threats listed above:

- Pre-configured passwords and, ideally, pre-configured usernames to be changed during initial setup of the application.
- Ensure the methodology of password retrievals is secure and does not provide any details of a valid account.
- Ensure the interface is not exposed to XSS, SQLi or CSR.
- Validate login credentials, and prevent them being shared over the network.
- Disallow use of weak passwords.
- Lock the account after a predefined number of failed login attempts.

2.3 Ensuring Data Integrity

Data Integrity: information integrity is the common characteristic of fulfillment and actualization of information. Information trustworthiness additionally represents the definiteness of information concerning administrative rights. It is left with the unauthorized access to the procedures, rules, and principles defined during the plan stage. At the point when the uprightness of information is secure, the data collected in a database will stay overall, precise, and reliable irrespective of to where it is stored and how frequently it is accessed. Information uprightness additionally assures that the information a user shares is protested from outside access.

Guaranteeing information integrity implies ensuring that information is unique, reliable, inferable and exact. Data integrity must be ensured at all phases of its life cycle; when it is made, transmitted, being used. Furthermore, there is no affirmation that the integrity of current information maintained.

One of the security models to provide data integrity is that of the CIA triad. Figure 2.2 lists the main building blocks of the model that revolve confidentially, integrity and availability. Here we discuss an overview of the CIA model.

With regards to IoT, confidentiality provides for securing and protecting of IoT gadgets, integrity cares for the information contained inside the gadget while availability covers accessibility of the gadget.

At the point when we consider information of an IoT gadget, we should consider not only the information being created or utilized by it, but also all parts of the programming code, arrangement parameters, and working framework of the code. To control the procedure of uprightness it is useful to consider three unique features that information can contain: to be specific, consistent and in process. Figure 2.2 shows the deterioration of the Integrity rule into sub-standards, lastly usage that the IoT gadget can fuse in securing the integrity of its information.

Any breach of information correctness defines that an IoT gadget can't work correctly, and additionally shows the gadget is being misused and is potentially a platform from which various assaults can be driven. The standard technique for confirming the integrity

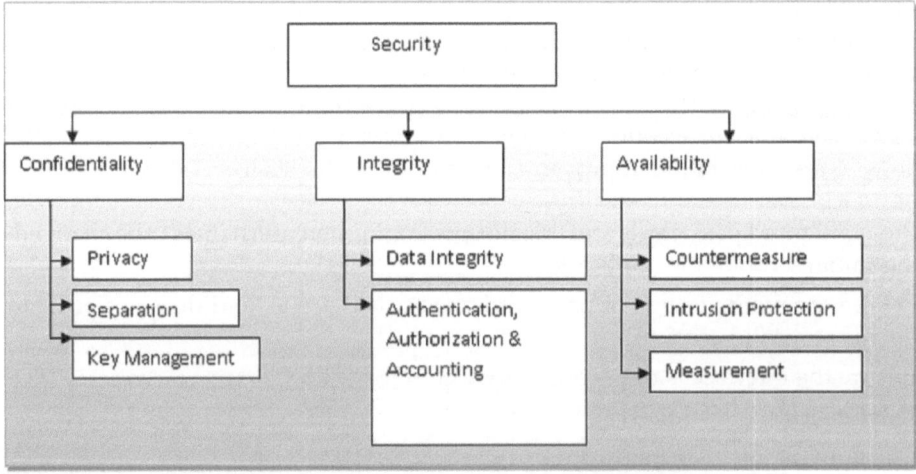

FIGURE 2.2
CIA model.

of information is by the mathematical method called "hashing", in which the secure hash algorithm (SHA) is the most often used one.

Integrity of data ensures that the data remains undamaged against malevolent attacks when transmitted over a network. Implementing unique enumeration values through the algorithm ensures that the APIs are called with criteria unique to that API. The CIA model defines a unique, simple and beneficial method to detect and represent the requirements for implementing security in the given IoT device. Maintaining data integrity is unquestionably one of the vital parts of security, and similar considerations and approaches should be made for that of secrecy and accessibility. Wind River give both a wide range and profound scope of items and arrangements that execute the integrity guideline of the CIA Triad for clients' IoT needs.

Data Integrity deals with the protection of data being shared and transferred over the network through the IoT devices. It is defined that the data shared between these IoT devices is strictly secured and the devices themselves are protected. The reason being that the data should be confidential as it contains personal information, financial transaction details, health data, and so on. There are certain security algorithms implemented which should be able to:

- encode/decode data;
- use a manipulative key;
- adapt with all types of devices;
- ensure secrecy & integrity of data/information.

With different cryptographic methodologies being implemented in different areas such as banks and biometrics, a few are discussed below:

2.3.1 Triple DES

Triple Data Encryption Standard (TDES) was made when the DES calculation was failing to keep up with the advancements. The point of making triple DES was to offer clients greater security as it would be a lot simpler utilizing various DES Composite Functions than to deal with new ciphers. TDES—otherwise called Triple Data Encryption Algorithm—is a symmetric key cryptography calculation where both the sender and collector utilize a similar key to play out the encryption/decryption process. TDES actualizes security by utilizing three cipher blocks and three keys in each data square.

2.3.2 AES

Hui Suo et al. [3] have defined AES is a subgroup of Rijndael Cipher that can procedure a block cipher of 128 bits (block length). AES calculations permit cipher keys of 128 (AES-128), 192 (AES-192) and 256 bits (AES-256). The calculation utilizes rounds to play out the encryption/unscrambling process and the number of rounds is reliant on the size of the key. For AES-128 it uses 10 rounds, for AES-192 it uses 12 rounds and AES-256 it uses 14 rounds. The essential unit of activity in AES calculations is a byte.

2.3.3 RSA

Hui Suo et. al. [3] in their work have stated RSA cryptosystem was structured as an after effect of the issue with key circulation in other prevailing cryptosystems. RSA cryptosystem is a sort of cryptosystem which takes out the requirement of conveying keys to beneficiaries to protect them over the transmission channel before transferring the first planned message. In RSA cryptosystem, the encoding key is open while the decoding key is private. This implies for decoding that only the individual with the matching decoding key can decode the message. The open key and private key are numerically related.

There are many considerations for ensuring data integrity for IoT devices including (but not limited to):

- Merchant/supplier appraisal.
- Verification and definition of the Entity Relationship (ER) model.
- Definition of the various security protocols implemented in IoT devices such as host identity protocol (HIP), host identity protocol pre-shared keys (HIP-PSK), contract network protocol (CNP).
- Defining and verifying the use of cryptography through key management protocol (KMP), elliptic curve Diffle hellman (ECDH) listed by Hui Suo et al. [3] for IoT communication.
- Defining methods for good data management like federation architecture, schema support, efficient indexing, and scalable archiving support.
- Identity and access management (IAM).

2.4 Insecure Network Services

Using insecure network services we can identify the vulnerabilities in the system that are utilized to access the IoT gadget that may permit an intruder unapproved access to the gadget or related information. Specific security vulnerabilities that could lead to this issue include:

2.4.1 Vulnerable Services

A vulnerability assessment is a way toward distinguishing, measuring, and organizing (or positioning) the vulnerabilities in a framework. Examples where vulnerability assessments are performed include, but are not restricted to: information technology systems, energy supply systems, water supply systems, transportation systems, and communication systems. Such assessments may be conducted on behalf of a range of different organizations, from small businesses up to large regional infrastructures such as IP Security Cameras, Virtual Assistants, HVAC systems, Smart TVs etc.

2.4.2 Buffer Overflow

A buffer overflow, or buffer overrun, is a typical application coding botch that an aggressor could adventure to access your framework. To adequately relieve buffer overflow vulnerabilities, it is essential to comprehend what buffer overflows are, what risks they pose to your applications, and what procedures aggressors may use to exploit these vulnerabilities.

2.4.3 Open Ports via UPnP

If you have at any point connected a USB cable to a PC, you've experienced the "attachment and play understanding", yet things are regularly not all that direct with organized gadgets. How does another printer, camera, espresso machine, or toy realize how to append to your system, and afterward arrange your switch to take into account port access?

UPnP is a helpful method for permitting devices to discover different gadgets on your system and if necessary adjust your switch to take into consideration gadget access from outside of your system. Through the Internet Gateway Device Protocol, a UPnP customer can get the outside IP address for your system and include new port sending mappings as a major aspect of its arrangement procedure. This is incredibly advantageous from a buyer point of view as it enormously diminishes the multifaceted nature of setting up new gadgets. Lamentably, with this comfort comes numerous vulnerabilities and a history of large-scale assaults which have misused UPnP.

2.4.4 Exploitable UDP Services

User Datagram Protocol (UDP) is a necessary convention that puts forth a best attempt to convey information to a remote host. In any case, on the grounds that the UDP convention is a connectionless convention, UDP datagrams sent to the remote endpoint are not ensured to show up, nor are they ensured to land in a similar arrangement in which they are sent. Applications that utilization UDP must be set up to deal with missing, copy, and out-of-succession datagrams. To forward datagram utilizing UDP, you should realize the system address of the gadget facilitating the administration you need and the UDP port number that the administration uses to impart.

There are several potential false positives and false negatives that can occur. For some systems, when performing outputs of numerous ports on numerous frameworks simultaneously, the genuine examining bundle probably won't make it to the objective host because of system congestion. Since no reaction is received, the scanner may respond that the port is open. In the event that no reaction is received, it should check once more, however this could likewise cause more bundles to be utilized in the test and influence an alternate test parcel to not arrive at its objective. The "U" in "UDP" is unreliable. Any switch or switch en route can drop the parcel on the off chance that it gets excessively occupied. Likewise intensifying the issue, firewall controls between the scanner and the objective may unequivocally deny outbound ICMP messages. For UDP parcels that advance toward a shut port and have the framework react with an ICMP port inaccessible message, this bundle may be dropped at the departure firewall or switch. Since this ICMP message isn't gotten, the scanner may inaccurately accept that everything is acceptable and the UDP port is to be sure open.

2.4.5 Denial of Service

A Denial of Service (DoS) attack listed by Sufian Hameed et. al. [7] takes place when the real/valid user is denied access to the services he/she requires via various gadgets, devices, platforms. The denial happens due to a malevolent attacker being present on the network. The area of influence includes emails, websites, or banking transactions, over the affected PC or system. The access to the service is denied by flooding the host with traffic until either crashing the host or stalling the access to the service. The DoS costs the host both monetary as well time wastage while the administrative rights are locked.

2.4.6 DoS via Network Device Fuzzing

A fuzzer is a code which infuses semi-arbitrary information into a program stack and distinguishes the intrude or attacker by Shafalika Vijayal et al. in their paper [6]. The data-generation part is made of generators, and vulnerabilities depend on troubleshooting instruments. Generators ordinarily use blends of static fluffing vectors (known-to-be-risky qualities) vectors by Angelo Furfaro et. al. [8], or absolutely irregular information. Recently created fuzzers utilize hereditary calculations to interface infused information. Such devices are not open yet.

A fuzzer will initiate attacks using:

- numbers (signed/unsigned integers/floating point)
- chars (URLs, command line inputs)
- metadata: user-input text (id3 tag)
- pure binary sequences

A simple method to use fuzzer is by putting it over various combinations to the list of values that prove to be dangerous.

- *In case of integers*: use zero, negative or large numbers
- *In case of characters*: use escaped, interpretable characters and instructions (for SQL requests, quotes/commands)
- *In case of binary*: use random combinations of zeroes and ones

There are many considerations that could help prevent the threats mentioned above.

- Minimize ports left free and available.
- Ascertain that network services are not prone to buffer overflow and fuzzer damage.
- Ensure that network services are not vulnerable to DoS attacks affecting the devices over the local and other networks.
- Ensure that the network ports are not open to UnPn through Internet as it allows unfair mapping.

2.5 Lack of Transport Encryption

Lack of transport encryption relates to the information being traded with the IoT gadget in a decoded manner. This could lead to an interloper getting a sniff of the information and either storing it for some future use or trading off the gadget itself.

Explicit security vulnerabilities that could implement this issue include unencrypted services via the Internet, unencrypted services via the local network, poorly implemented SSL/TLS, misconfigured SSL/TLS.

2.5.1 Unencrypted Services via the Internet

On decoded web, one of the severest perils is that your own credentials fall into the hands of individuals you don't want them to.

Examples of unencrypted Internet services are Telnet (replaced by SSH), FTP (still in use, but on the way to becoming SFTP or SCP), and traditional protocol on HTTP (on the way to getting replaced by HTTPS). Encryption comes with a cost, and it's not always easy to implement.

MSPs have a significant task to carry out in helping their customers address these encryption challenges. Effectively doing so won't just assist customers with securing their information, however; it will likewise diminish the expenses and harm when breaks unavoidably happen.

How are MSPs helping? Methods for this include:

- To help the clients carry out security checkups. Audits by IT teams are required to check the performance of the software internally to identify potential vulnerabilities and also define which data is to be encrypted and which method is best suited.

- Provide clients with the encryption techniques required. This will be varying by platform of implementation (medical aid, for instance, has strict and well drafted requirements) and could be affected by the type of remote services you already provide.

- Depending upon the business strategies, draft the encryption strategies to offer assistance for implementing technology to bring about those polices via training, automation, and control and management tools.

- Implement the available encryption techniques into the service tool sets. Barracuda Essentials, for example, provides outbound email protection using encryption.

- Implement security data to analyze the encryption technique, and provide clients with methods to identify their most vulnerable data section. For example, Barracuda's Event Log Analyzer for the Web Application Firewall automatically collects and analyzes data and creates ready-made (and easy-to-understand) reports on user activity, attack mitigation and other security incidents.

2.5.2 Unencrypted Services via the Local Network

When data is shared through the IoT devices in an unencrypted form, it becomes easier for hackers to gain access to that data. Likely using unencrypted services through web, local network, poorly installed SSL/TLS stated by Hui Suo et. al. in their work [3], misconfigured SSL/TLS and lack of firewall can prompt this issue.

2.5.3 Poorly Implemented SSL/TLS

Unfortunately, SSL/TLS is difficult to revise protocol for, because:

- Being a protocol that provides a secure end-to-end encryption which makes it difficult for an eavesdropper to check data, it does not ensure security of data for a misconfigured server which falls due to less protection from modern threats.

- Many configurations are unstructured.

- There are a lot of methods available which every now and again simply work around the issue by affecting the security of the target.

2.5.4 Misconfigured SSL/TLS

- Bad ciphers, like RC4-SHA are the ones allowed by the server. Clients like curl 7.35.0 have disabled these ciphers by default and there are recommendations for others like Microsoft Windows.
- Rather than revising the protocols, the administrators have tried to make systems safe against POODLE[1] by disabling all SSL 3.0 ciphers. A maximum of two clients can be connected by these ciphers, as these are required by TLS1.0 and TLS1.1.

There are many considerations that may protect from the threats mentioned above:

- Assure that data is encrypted by using protocols such as SSL and TLS while transiting over the networks.
- Assure that other encryption methods are available in the absence of SSL or TLS.
- Assure that only trusted encryption protocols and techniques are implemented.

2.6 Privacy Concerns

Security concerns arise by the assortment of individual information in the absence of legitimate assurance of that information defined by Mookyu Park et al. [4]. Protection concerns are anything but difficult to find by just inspecting the information that is being gathered as the client sets up and enacts the gadget.

2.6.1 Security Risks

- IoT gadgets are related to your work area or PC. Absence of security increases danger for your own data while the information is gathered and transferred to the IoT gadget.
- IoT devices are connected with a consumer network. So if the IoT device contains any security vulnerabilities, it can be harmful to the consumer's network. This vulnerability can attack other systems and even destroy them.
- Physical security defined by David Hutter (2016). *Physical Security and why is it important* [1] is risked by unauthorized people exploiting the vulnerabilities.
- Because IoT gadgets are associated with the Internet, they are powerless against similar sorts of cyber threats that can harass shopper, business, modern, and administrative PC frameworks. In September 2016, the "Mirai botnet" exploited the IoT devices at a large scale by controlling thousands of IoT devices to launch the Distributed DoS attacks by shutting down the target websites. Since IoT gadgets depend on availability to work, they make a typical assault vector for programmers to access a whole system. Numerous IoT gadgets are designed with fundamental patterns of programming, and are much of the time not structured in light of cyber security, which builds the dangers they present.
- The updates are meant to protect a system against security breaches. But often these vulnerabilities are left unaddressed for a long time. On behalf of IoT gadgets with

usually long spans of usability there is a drawback that the manufacturer will stop support, or simply cease trading.

- There are additional special security dangers presented by IoT gadgets' utilization of cloud administrations. Putting away information on remote servers fundamentally expands the likelihood that the information will be compromised. Parting power over the gadget and the information decreases the capacity of any one supplier to restrain get to, and reliable security gets subject to harmonization of information security rehearses among the different gatherings liable for its assortment, transmission, and capacity. The most encouraging reaction to the expanding unpredictability of these frameworks would be a boundless selection of a solitary, steady arrangement of principles. The NIST Cybersecurity Framework being one of the most important norms at the government level was as of late updated in January 2017.

- A poor security standard can lead to serious results such as physical damage and harm. Some such cases can be listed as the hacking of a car by an online malevolent attacker, which could initiate vehicular crime by giving the access and control of the fundamental elements of a vehicle, including its brakes. Another category of IoT device that could be hacked with server consequences is personal medical devices, such as defibrillators, pacemakers, and insulin pumps; hacking of any of these devices could lead to human physical injury or death.

2.6.2 Privacy Risks

- In IoT, gadgets are interconnections of different equipment and programming, which have clear odds of delicate data spilling through unapproved control.

- All the gadgets are transmitting the client's private data, for example: name, address, date of birth, health information, and Visa detail; and significantly more without encryption.

- If a malevolent attacker hacks into a cell phone or PC, there is a high probability that the control of the device may be lost remotely and left undetected, leading to loss and/or exposure of the victim's data, financial circumstances, and emails. As network of the IoT devices increases with smart homes, smart vehicles. This information leakage can be due to the regular illustrations of the users through the exposure to various tasks on the Web. All such activities are monitored by organizations for providing a better user experience.

- By reading the policies that companies provide, consumers can become aware of how companies access and control their data. This means that companies must bring changes to the policies by making consumer aware of the privacy.

The aggressor can extricate the private data of the client through the information gathered and data spilled. The aggressor can find the security by utilizing the following stated dangers:

- *Keystroke inference attack*: such attacks affect not only the device but also devices placed near to the target. This attack uses the input devices like touchpad and keyboard to determine the username and passwords entered by the user. The data is fetched by the deviation caused on the sensors by the change in direction of the device.

- *Task inference attack*: This type of attack is used to find out the information regarding the current tasks being carried out on a user's smart device so that the device's state can be duplicated. Such an attack helps find out which applications are being run on a user's device connected over the network.

- *Eavesdropping*: Some voice applications use a malicious program installed to fetch the content of conversation extracted from audio sensors installed in AI speakers without the knowledge of the user. A malware that incorporates the voice assistant application can use a range of malicious activities such as voice duplication to carry out financial fraud through phone.

- *Location inference attack*: This attack uses a side-channel to control the IoT devices to find the personal information such as home or work address.

Some of the Information assets that cause various scenarios of privacy breach include:

- User Credentials have privacy concern over typing username and passwords to touch screens and pads by having motion sensor running at backend to define keystroke inference and eavesdropping attacks.

- Mobile personal data has privacy risks such as the leaking of information by the camera sensor, tapping of audio through sensors, etc. to support keystroke inference, eavesdropping attacks, location inference and task inference attacks.

- Home, as an asset, can confirm to the user, through environmental and magnetic sensors, the position of that user through location inference.

- Home structure as an asset can check the shape of the house through the help of camera sensors and internal layout of the house through wavelengths made familiar by audio and light sensors to define the eavesdropping and task inference.

There are many considerations that prevent against the threats mentioned above.

- Assuring that only a specific data that defines the basic functionality of device is made available.
- Assure that only relevant data is made available.
- Guarantee that any information gathered is re-recognized.
- Assuring that the data collected is encoded in a proper fashion.
- Assuring the personal information of the client is preserved.
- Assuring that the personal data is accessible by only the authorized people.
- Ensure that a time limit is set for retaining the data acquired.
- "Notice and Choice" should be ensured at the end if data collected is more than what was required.

2.7 Insecure Software/Firmware

The absence of capacity for a gadget to be updated presents a security breach. Gadgets ought to be able to be updated when vulnerabilities are found, and programming/firmware updates can be unreliable at times. Firmware contain crucial system information through which attacker can get control of a system during the system startup and boot process and they can completely "own" what happens with that system. The gadgets remain vulnerable to the updates being generated by the companies to address various issues in order to deal inclusively with the security issues of the software/firmware. Furthermore, in the event that the gadgets have hardcoded delicate accreditations, in the event that these certifications get uncovered, at that point they remain so for an unknown and indeterminate timeframe. Explicit security breaches that can generate this issue include this issue include:

- Updates are left unencrypted when fetched.
- The files that conduct updates are not properly encrypted.
- All the updates that go unverified before download.
- Sensitive/critical information being part of firmware.
- There should be no functionality of obvious updates.

Software engineers could better explore deficiencies in IoT security to ambush the devices themselves, as an entry point for a wide scope of misconduct, which could consolidate DoS attacks, malware dispersion, spamming and phishing and credit card robbery, etc. In this way, before a gadget hack prompts income theft or harm to your organization's reputation, or something even more undesirable, it is essential to know about the eight most basic firmware vulnerabilities to ensure you haven't left the front entryway open to your system:

1. *Unauthenticated access*: It is common of all the breaches to seek access of firmware by the malevolent actors in order to take control of the IoT device for the purpose of gaining access to crucial data and information.
2. *Weak authentication*: A poor authentication technique that includes bad cryptographic algorithm being exploited by the brute-force attacks leads to easy access of devices over the network by malevolent actors guessing passwords.
3. *Hidden backdoors*: To provide easy access to the firmware, hidden backdoors are a hacker's easy method to breach, and inject code, etc. into an embedded device to give remote access to anyone with the "secret" authentication details. Backdoors prove to be helpful to customer support when exposed to malevolent users. It is very easy for hackers to recognize the backdoors.
4. *Password hashes*: Users retain default passwords or do not change them for the firmware. Both the situations lead to devices prone to breaches which can be exploited. Both result in IoT devices being relatively easy to exploit with the execution of a DDoS attack. One such Incident happened when Mirai Botnet affected millions of devices connected over the Internet to take down Netflix, Amazon, and The New York Times, among others in 2016.
5. *Encryption keys*: When stored in a format that can be easily hacked, like variations of the DES, first introduced in the 1970s, encryption keys can present a huge problem for IoT security. Even though DES has been proven to be inadequate, it's still in use today. Programmers can misuse encryption keys to listen to correspondence, access the gadget, or even make rebel gadgets that can perform malignant acts.

6. *Buffer overflows*: When coding firmware, issues can rise if the engineer uses unstable string-dealing with limits, which can incite buffer overflows. Assailants invest a great deal of energy taking a look at the code inside a gadget's application software, attempting to make sense of how to cause inconsistent application conduct or crashes that can open a way to a security break. Buffer overflows can allow software engineers to remotely find a good pace that can be weaponized to conduct DoS and code injection assaults.

7. *Open source code*: Open source stages and libraries engage into the snappy improvement of current IoT things. In any case, on the grounds that IoT gadgets every now and again utilize outsider, open source parts, which ordinarily have obscure or undocumented sources, firmware is normally left as an unprotected assault surface that is overpowering to programmers.

8. *Debugging services*: Debugging data in beta renditions of IoT gadgets provides attackers with inside setup of a gadget. Lamentably, troubleshooting frameworks are regularly left underway, giving programmers access to the equivalent inside information on a gadget.

There are many considerations that prevent the threats mentioned above:

- Ascertain that the device has the capability to upgrade (most important).
- Make sure that the updated file is encrypted using applicable encryption technique.
- Assure that the upgraded file is transported through an encrypted network.
- Ascertain that sensitive data is not published over the network.
- Ensuring that the upgrade is validated and authenticated before allowing the update to be uploaded and transferred.
- Ascertain that the updated server is secure enough to deal with various vulnerabilities.

2.8 Poor Physical Security

Physical security weaknesses are accessible when an attacker can disassemble a device to easily find a workable pace medium and any data set on that medium. Shortcomings likewise exist when USB ports or other external ports can be utilized to get to the gadget utilizing highlights proposed for arrangement or up keep. This can cause unauthorized permission to the gadget or the information. Specific security breaches that could generate this issue include:

- Clearing the storage media.
- Permission to access the software via USB ports.

When we discuss the significance of cybersecurity, we regularly discuss the Internet and the Cloud. We comprehend the requirement for solid client validation, occasion observing, movement logging, and encryption of information and the entirety of the commands that should be worked in to our IT systems to keep us safe. As the IoT keeps on developing (expected to comprise 20–50 billion gadgets by 2020), we have to ensure our physical security is taken a gander at with similar eyes as well.

- Physical security is the insurance of life and property and incorporates things as different as individuals, equipment, firmware, and even the information that from an occasion that causes misfortune or harm. Access control and video surveillance are used to perform specific security capacities: watch out for the malevolent users and don't let them gain access. CCTV frameworks have cameras connected through network link with restrictive correspondences to a video controller that has its repository at some place. Clients hold certificates like identifications, tokens that are directly associated to gain the access of the broad device framework.

- A breach of the free physical security arrangement could be completed with practically zero specialized information by the assailant. Also, cataclysmic events and mishaps are an unavoidable piece of our day by day lives. Be that as it may, the IoT and interconnectivity is largely affecting the physical security industry. This carries two inquiries to mind; how would you interface physical security gadgets to the Internet and guarantee they were protected from intruders, and how might you utilize your present observation, to control and interrupt the location of gadgets set up?

- The IoT offers an approach to improve our physical security and access control frameworks. With the IoT and cloud the executives, we can utilize our current arrangements however enhance them with adaptability and framework changes that are shared over the system continuously. Frameworks and updates should be publicly released and accessible on numerous platforms, much the same as in organized innovation. Producers should make sense of how to serve this market. What's more, numerous customary systems administration sellers are as of now taking a shot at open camera IP stages that empower connection to edge-based capacity and offer an API for application improvement.

- What occurs if the Internet crashes? As the IoT devours an ever-increasing number of gadgets it turns out to be similarly as effective as force. All frameworks need excess and safe-check components. We can't depend on a solitary purpose of disappointment and need to consider the things that can turn out badly and put resources into these advancements.

IoT exists, and it is setting down profound roots. By 2020, Gartner predicted, the IoT will be include at least 26 billion "units." The specific controls referenced above will come to nothing if the producers of IoT contraptions don't think about them. The activities are straightforward as noted beneath:

- A security check of the IoT devices must be conducted to check upon the vulnerabilities.

- As per above, guaranteeing the security is a vital piece of the device lifecycle enhancement with the goal that it gets implanted into the gadget and not as an idea in retrospect. The messages seem to be quite easy for the enterprises.

Take into consideration the following:

- Determine your basic data resources and disengage/ensure those. Conventional security methods here are as yet compelling.

- Imagine that you are getting busted! You must be prepared with techniques to deal with this situation at any point in time.

How might I secure my gadgets? Given the idea of numerous IoT gadgets it is difficult to secure against a determined assailant. In any case, there are scopes of measures that can be taken to secure against most situations and we'll take a look at some of these. While evaluating safety efforts for a gadget, consider its excursion from when it leaves the production line, goes from there to retail, and afterward the customer, at that point throughout everyday life and as far as possible of its life. Different protection methods can be utilized, yet the cost model will certainly be a characterizing factor with respect to how a lot of financing is accessible to relieve threats. The IoTSF article on Classification of Data talks increasingly about security hazards, security financing. So the initial phase in protecting a gadget is to expel all physical, radio or optical ports that were involved in development purposes. Obviously at any rate one port is required to interface connection of the gadget to the local network or into the Cloud, however all other ports that are never again required ought to be evacuated, including any circuit board tracks that associate the port into the hardware. So also, all test points need to be expelled, including pins and circuit tracks, or if nothing else successfully impairing test access by for example blowing on-chip wires for JTAG.

On the premise that in designing the gadget there must have been some type of authorized port, this will be a practical objective of consideration by attackers. Accordingly, authorized access to this port must be secured appropriately. The organization administration must utilize a protected convention, for example, SSH; uncovering an unreliable convention, for example, Telnet on an organization port successfully leaves the gadget fully open. The administration of certifications to get to this port must be secure and successfully figured out how to guarantee get to can't be undermined. The IoTSF article on Credential Management speaks progressively about procedures for overseeing certifications safely. Chips that manage critical capacities are now and again expelled from circuit sheets by attackers, so they can interface the chips to test hardware for investigation. To forestall this, key chips can be epoxied to the circuit board or different highlights with the end goal that the chip gets wrecked during the time spent evacuating it. Additionally, the whole hardware can be inserted in a square of tar, which will render it difficult to reach by anyone except the most determined.

Obviously, the last line of protection is simply the gadget packaging. This can give physical insurance to the innards, structure some portion of the mounting, and help shield against recognition of radiations. One alternative can be to render the gadget permanently disabled if the case is opened. There is consistently a theft that a gadget may get messed with some place in the production network, maybe in a distribution center or in travel. Anti-tamper packaging or seals can be applied to give a visual sign of any endeavor on meddling with the item. So, there are numerous ways a gadget can be ensured against noxious acts, yet much relies on utilization situations, specialized ability, and especially on the cost model; in other words, what level of assurance can be realistically managed. We know that IoT Gadgets are a means of communication over the network therefore the streams of data to transferred are to be carefully monitored. IoT contraptions associated over the Web can prompt a lot of simple life.

There are many considerations that can prevent the threats mentioned above:

- Assure that the data storage medium cannot be easily removed.
- Ascertain that the stored data is encrypted.
- Ensure that any malicious activity cannot happen through USB or other external ports.
- Ensure that the IoT gadgets cannot be disassembled.
- Minimize the use of external ports, such as USB, in the functioning of the product.
- Ascertain that the administrative capabilities are limited by the device.

2.9 Conclusion

The vulnerabilities and security issues related with IoT can be radically decreased by executing security examination. These include gathering, associating, and investigating information from numerous sources that can help IoT security suppliers to recognize potential dangers and check such risk from the beginning. In this manner, there is a requirement for multi-dimensional security investigation separated from checking IoT doors alone. IoT is a large number of heterogeneous gadgets related into one framework associated by means of radio signals. IoT manages enormous volumes of information. Along these lines, it is exceptionally simple to perform attacks on it. To deal with this issue, different machine learning methods and AI techniques are in use. As intricacy of the framework is so high, accessing the methods of interruption requires complex computations to take care of the issue in effective manner. To adapt to different situations, we have proposed a review on various techniques and algorithms to address irregularities and intrusions in associated devices.

Note

1 Padding Oracle On Downgraded Legacy Encryption.

References

Abla El Bekkali, Mohammed Boulmalf, Mohammad Essaaidi, Ghita Mezzour, Securing the Internet of Things (IoT): Systematic Literature Review, *6th International Conference on Wireless Networks and Mobile Communications (WINCOM)*, Marrakesh, Morocco, 2018. 10.1109/WINCOM.2018.8629652.

Sufian Hameed, Faraz Idris Khan, Bilal Hameed, Understanding Security Requirements and Challenges in Internet of Things (IoT): A Review, *Journal of Computer Networks and Communications*, Article ID 9629381, 2019.

David Hutter, Physical Security and Why It Is Important, SANS Institute Information Security Reading Room, June 2016. Available at: https://www.sans.org/reading-room/whitepapers/physical/physical-security-important-37120

Mohit Mittal, Shafalika Vijayal, Detection of Attacks in IoT Based on Ontology Using SPARQL, *7th International Conference on Communication Systems and Network Technologies (CSNT)*. doi: 10.1109

Mookyu Park, Heangrok Oh, Kyungho Lee, Security Risk Measurement for Information Leakage in IoT- Based Smart Homes from a Situational Awareness Perspective, Published online 2019, May 9. doi: 10.3390/s19092148

Hui Suo, Jiafu Wan, Caifeng Zou, Jianqi Liu, Security in the Internet of Things: A Review, *International Conference on Computer Science and Electronics Engineering*, 2012.

Shafalika Vijayal, Mohit Mittal, *Intrusion Detection in IoT based on Neuro-Fuzzy Approach*, IJSRCSEIT, Vol. 2, Technoscience Academy, Jammu & Kashmir, India, 2017.

3

Physical Layer Security for Energy-Efficient IoT with Information and Power Transfer

Aaqib Bulla and Shahid M. Shah

CONTENTS

3.1 Introduction

Internet of Things (IoT) is a giant integration of billions of physical devices and people over the internet. With device-to-device and device-to-human communication capabilities, IoT has evolved from aspirational visions to real-world applications because of the massive technological developments in electronic devices, micro-electromechanical systems, and wireless communications, especially 5G technologies. The collection and sharing of data over a billion connected devices at very high data rates have posed several challenges in the context of energy consumption and security. Energy-efficient wireless communication has been a research focus for quite a while now, and several solutions including Advanced Physical layer Techniques, Network Architecture Designs, Resource Management Schemes and specifically different Deployment Schemes in IoT have been proposed for improving the Energy Efficiency (EE) in wireless communication networks (Jiang, C. and Cimini, L.J., 2013), (Huang, J., Meng, Y. et al., 2014), (Huang, J. et al., 2017a). DAS, a key technology well known for expanding network coverage and increasing data rates is being widely studied in the field of energy-efficient wireless communication (Choi, W. and Andrews, J.G., 2007), (Lee, S.R. et al., 2011). Specifically, EE optimization in DAS was considered in (Zhang, J. and Wang, Y., 2010), (Chen, X., Xu, X. and Tao, X., 2012), (He, C., Sheng, B., Zhu, P. and You, X., 2012), (Kim, H., Lee, S.R., Song, C., Lee, K.J. and Lee, I., 2014). While the DAS is known for enhancing the system performance, it is going to play a significant role in wireless power transmission due to the reduced transmitter-receiver access distance. This technology of Simultaneous Wireless Information and Power Transfer (SWIPT) is being studied as a promising solution for sustainable operation of IoT devices in energy-constrained wireless communication networks (Huang, J., Xing, C.C. and Wang,

C., 2017b), (Jameel, F., Haider, M.A.A. and Butt, A.A., 2017), (Perera, T.D.P., Jayakody, D.N.K., Sharma, S.K., Chatzinotas, S. and Li, J., 2017), (Ng, D.W.K., Lo, E.S. and Schober, R., 2013), (Guo, S., Wang, F., Yang, Y. and Xiao, B., 2015). Recently, EE optimization in the context of SWIPT based DAS was discussed by Huang, Y., Liu, M. and Liu, Y. (2018). While SWIPT promises a significant reduction in energy consumption by reducing frequent battery charging or replacement, security is still a matter of concern for a billion device based IoT network.

Recently, security in an IoT network in addition to energy efficiency was studied in (Oliveira, D., Gomes, T. and Pinto, S., 2018), (Zhao, Y.S. and Chao, H.C., 2018), (Ahuja, B., Mishra, D. and Bose, R., 2019). Due to the broadcast nature of wireless channels, information signals are susceptible to eavesdropping resulting in potential information leakage or even EH by eavesdropping nodes in SWIPT environment. The problem of information security was conventionally dealt with cryptographic algorithms at the higher logical levels. However, in IoT, we usually have lower power and less complex devices like sensor nodes, which may not have such levels of computational capability. Therefore, much attention is given to the data transmission security at the physical layer, where several aspects of the radio channel are exploited to ensure communication security (Bloch, M., Barros, J., Rodrigues, M.R. and McLaughlin, S.W., 2008), (Gopala, P.K., Lai, L. and El Gamal, H., 2008), (Mukherjee, A., Fakoorian, S.A.A., Huang, J. and Swindlehurst, A.L., 2014). The fundamental essence of security at the physical layer is information-theoretic characterizations of secrecy capacity by Aaron Wyner in context of wire-tap channels (Wyner, A.D., 1975).

Secrecy of data transmission from an information-theoretic perspective was first studied by Claude Shannon (Shannon, C.E., 1949). He considered a noiseless cipher system as shown in Figure 3.1. The transmitter transmits a message M to a legitimate receiver (Bob) while ensuring its secrecy from an eavesdropper (Eve) by sharing a secret key K with Bob that is unknown to Eve. To ensure the secrecy of the message M it is encrypted into a codeword X using the secret key K before being transmitted.

This idea of secrecy was generalized by Wyner's wire-tap channel model for a practical noisy communication channel, as shown in Figure 3.2, with no need of sharing a secret key between transmitter and legitimate user.

Authors in (Leung-Yan-Cheong, S. and Hellman, M., 1978) extended this result to Additive White Gaussian Noise (AWGN) wire-tap channel. With the aim that channel output at Eve should not reveal any information, while the Bob should recover the transmitted message reliably; the idea of secrecy capacity was used, which characterizes the

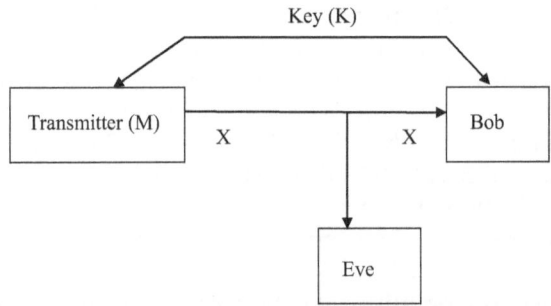

FIGURE 3.1
Shannon's noiseless cipher system.

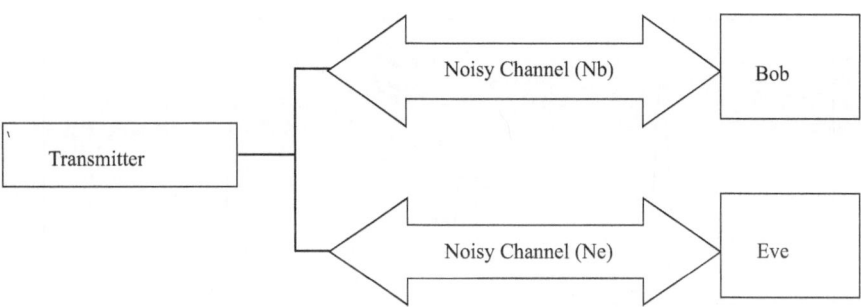

FIGURE 3.2
Wyner's wire-tap channel model.

maximum data rate at which both requirements are met. Mathematically, Secrecy capacity is the difference between the maximum achievable data rates over the main channel (transmitter-legitimate receiver) and the wire-tap channel (transmitter-eavesdropper). Moreover, in SWIPT systems, artificial noise can be used to ensure information secrecy while the wireless power transfer requirements are still met (Xing, H., Liu, L. and Zhang, R., 2015), (Yu, H., Guo, S., Yang, Y. and Xiao, B., 2017), (Jameel, F., Wyne, S., Junaid Nawaz, S., Ahmed, J. and Cumanan, K., 2018).

In this chapter, Security and Energy Efficiency optimization for wireless information and power transfer in a DAS based IoT network is discussed. The term 'Security' corresponds to the secrecy rate metric as defined above, where it is assumed that the transmitter-legitimate receiver channel is less noisy than the transmitter-eavesdropper channel. This secrecy rate metric is used to define SEE as a ratio of the secrecy capacity to the total power consumed at DA ports. The primary objective is to achieve the data rates which are beyond the Shannon capacity of the eavesdropper, making it incapable of decoding any information over the intercepted channel. Also, it has to be ensured that we do so at minimum possible energy consumption.

First, in Section 9.2, we consider the case of information transfer only and then the approach is extended to simultaneous information and power transfer in Section 9.3. In Section 9.4, a generalized system model is formulated for a practical IoT scenario.

3.2 Secure and Energy-Efficient Information Transfer

Let's consider a single user (Bob) being overheard by a single eavesdropper (Eve) in a DAS with N-Distributed Antenna (DA) ports, each one having a single antenna. As an example, a three DA port system is shown in Figure 3.3. A central control unit coordinates the overall operation of the distributed system, and all the ports are connected to each other and to the central control unit over dedicated channels (usually optical links).

The corresponding channel outputs at Bob and Eve are mathematically given by equations (3.1) and (3.2) respectively; refer to Appendix A for details.

$$y_b = \sum_{i=1}^{N} \sqrt{P_i S_i^{(b)}} h_i x_i + z_b \qquad (3.1)$$

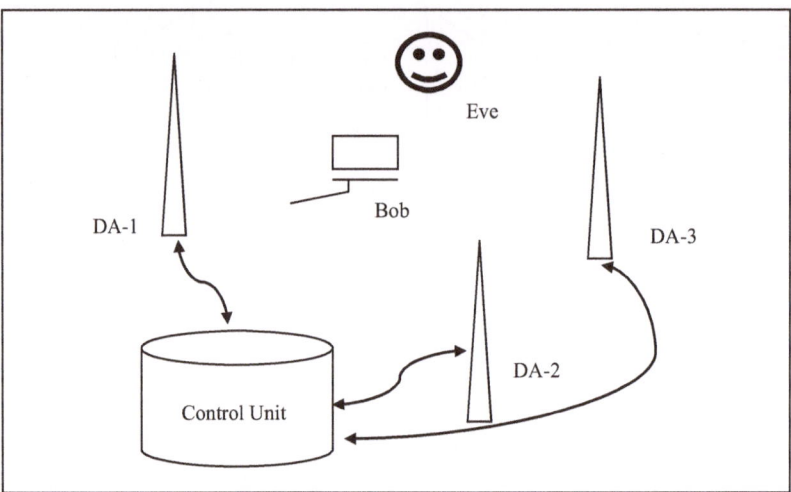

FIGURE 3.3
A Distributed antenna system with 3 DA ports.

$$y_e = \sum_{i=1}^{N} \sqrt{P_i S_i^{(e)}} \, g_i x_i + z_e \tag{3.2}$$

Where:

- P_i is the power consumed by i^{th} DA port,
- $S_i^{(b)} = \left(d_i^{(b)}\right)^{-\alpha}$ and $S_i^{(e)} = \left(d_i^{(e)}\right)^{-\alpha}$ are respectively the propagation path-losses for Bob and Eve with path-loss exponent α,
- h_i and g_i are the corresponding (circularly symmetric complex Gaussian) channel coefficients, which are assumed to be independent and identically distributed (iid) with zero mean and unit variance,
- x_i denotes the transmitted symbol from the ith DA port with average power $E[\,|x_i|^2]$, and
- z_b & z_e represent the AWGN processes with zero mean and variance σ_b^2 and σ_e^2.

Also, symbols transmitted from each DA port are assumed to be statistically independent. For the case of Information Transfer (IT), the receiver only decodes the information from the received radio signal, therefore the maximum possible data rate for Bob in this scenario is:

$$R_b = \log_2\left(1 + \sum_{i=1}^{N} \gamma_i^{(b)} P_i\right)$$

where $\Gamma_i^b = \dfrac{S_i^b |h_i|^2}{\sigma_b^2}$ is the effective channel gain to noise power ratio (CGNR) from the ith DA port to Bob. Now, in the presence of Eve, the transmitter at each DA port uses Wyner's wire-tap coding so as to transmit the data at a rate which exceeds the information capacity of Eve's channel. The achievable secrecy capacity R_S for Bob in this scenario is expressed as:

$$R_s = \left[C \left(\sum_{i=1}^{N} \Gamma_i^b P_i \right) - C \left(\sum_{j=1}^{N} \Gamma_j^e P_j \right) \right]^+ \tag{3.3}$$

where $C(x) = 1 + \log(x)$, $[x]^+ = \max(0, x)$ and $\gamma_i^{(e)} = \dfrac{S_i^{(e)} |g_i|^2}{\sigma_e^2}$ represents the corresponding CGNR from the jth DA port to Eve. (An outline of the proof of secrecy rate is provided in Appendix B.)

For an N-port DAS, we use the expression for secrecy rate given in (3) to define the Secure Energy Efficiency as:

$$\theta_{SEE} = \frac{R_s}{\sum_{i=1}^{N} P_i + P_c}$$

where P_c represents the total power consumed at the transmitter during various signal processing operations. Our main objective is to achieve higher secrecy data rate at minimum possible energy consumption, which in other words implies maximizing θ_{SEE} over different possible values of transmit power at DAS ports. Thus, having defined the objective function $\{\theta_{SEE}\}$, we formulate a constrained optimization problem as:

$$\max_{\{P_i\}} \theta_{SEE} \; s.t : 0 \le P_i \le P_{max,i} \tag{3.4}$$

To solve the problem given in (3.4), we derive the KKT conditions. (Refer to Appendix C for details.) In the optimization theory, KKT conditions are the necessary conditions for a general non-linear programming problem to be optimal if the regularity conditions are satisfied (Boyd, S., Boyd, S.P. and Vandenberghe, L., 2004).

Now, let's assume that the channel between Transmitter and Bob is always less noisy than the channel between Transmitter and Eve. In that case, θ_{SEE} is a quasiconcave function of $P_i (\ge 0)$ and hence any maximizer that exists will be optimal. The justification for this statement is given in the following Lemma.

Lemma 1: For $k > l$ and k, l, m and t all non-negative real numbers

$$f(t) = \frac{\log_2 (1 + kt) - \log_2 (1 + lt)}{t + m}$$

is a quasiconcave function of t.

Proof:
Let $g(t) = \log_2(1 + kt) - \log_2(1 + lt)$
Differentiating $g(t)$ twice with respect to t, we get:

$$g'(t) = \frac{k}{(1 + kt)} - \frac{l}{(1 + lt)}$$

$$g''(t) = \left[-\frac{k^2}{(1 + kt)^2} + \frac{l^2}{(1 + lt)^2} \right]$$

Clearly $g''(t)$ is negative as long as $k > l$, which means g is a concave function. Also, $h(t) = t + m$ is an affine function. Hence the given function $f(t) = \dfrac{g(t)}{h(t)}$, being the ratio of a concave and an affine function is a quasiconcave function. So, if x^* is a strict local máxima, then it is also global (Zappone, A. and Jorswieck, E., 2015).

Now, for the given optimization problem in (3.4), the Lagrangian function is:

$$L(P_i, \lambda_i, v_i) = \frac{R_s}{\sum_{j=1}^{N} P_j + P_c} + \sum_{j=1}^{N} \lambda_j P_j + \sum_{j=1}^{N} v_j \left(P_{max,j} - P_j \right)$$

Where, λ_i and v_i are the Lagrange multipliers corresponding to the constraints $P_i \geq 0$ and $P_i \leq P_{max,i}$ for $i = 1,\ldots,N$.

As per the KKT conditions, the optimal values $\{\lambda_i^*, P_i^*, v_i^*\}$ will satisfy the following set equations [28]:

$$\frac{\partial L}{\partial P_i} = f_i \left(P_1^*, P_2^*, \ldots P_N^* \right) + \lambda_i^* - v_i^* = 0$$

$$\lambda_i^* P_i^* = v_i^* \left(P_{max,i} - P_i^* \right) = 0, i = 1,\ldots, N$$

While, $P_i^* \geq 0$ and $\left(P_{max,i} - P_i^* \right) \geq 0$ and where:

$$f_i = \frac{\partial L}{\partial P_i} = -\frac{R_s}{\left(P' \right)^2} + \frac{1}{\left(P' \right)} \left\{ \frac{\Gamma_i^b}{C(B)} - \frac{\Gamma_i^e}{C(E)} \right\},$$

$$P' = \sum_{i=1}^{N} P_i + P_c, B = \sum_{j=1}^{N} \Gamma_j^b P_j, E = \sum_{j=1}^{N} \Gamma_j^e P_j$$

Now, since the objective function is twice differentiable, we use Sequential Quadratic Programming (SQP) to solve KKT conditions (Nocedal, J. and Wright, S.J., 2006), and as proved in Lemma 1, the function is quasiconcave for $\Gamma_i^b > \Gamma_i^e$; hence, the obtained solution is optimal.

In Figure 3.4, for $N = 5$ (DA Ports), different values of EE are plotted with respect to the maximum transmit power constraint. It is observed that energy efficiency gradually improves up to a certain point, and then saturates for higher values of P_{max}. Also, for the same P_{max}, SEE is less than EE, which implies there is a trade-off between secrecy and energy efficiency.

However, we can compensate for the reduction in energy efficiency by increasing the number of DA ports (N) as shown in Figure 3.5. The gain in energy efficiency with an increase in the number of DA ports (N) is due the reduced transmitter-receiver distance.

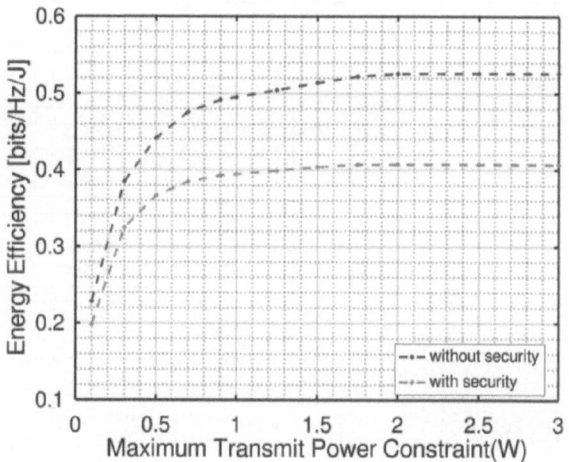

FIGURE 3.4
Energy efficiency with $N = 5$ and $P_c = 1W$.

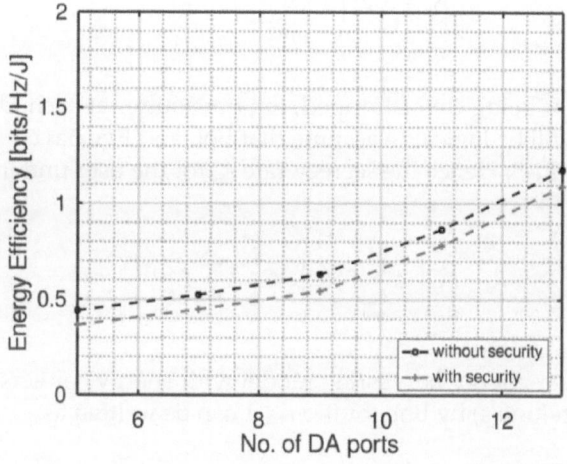

FIGURE 3.5
Energy efficiency with $P_{max} = 0.5\ W$.

3.3 Secure and Energy-Efficient Information and Power Transfer

Now, let's consider the case of wireless power transfer alongside information transfer over the radio link for a single user and single eavesdropper scenario. In SWIPT systems, as shown in Figure 3.6, a power splitter at the receiver of an EH device splits the signal power into two parts, with $0 < \Delta < 1$ part for Information Decoding and $1 - \Delta$ part for EH (E_h).

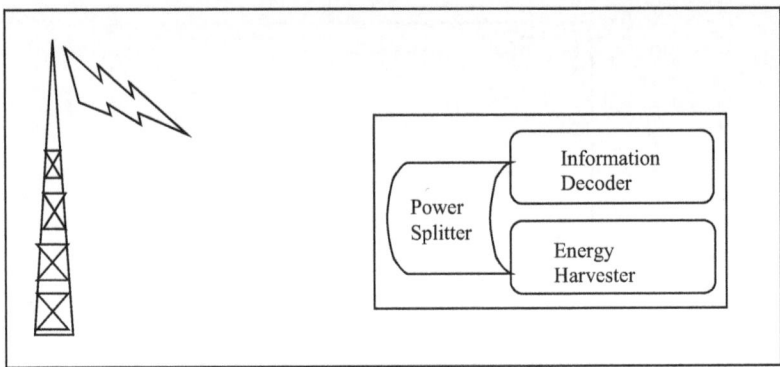

FIGURE 3.6
Receiver module in a SWIPT system.

As discussed in (Huang, Y., Liu, M. and Liu, Y., 2018), the maximum achievable rate for Bob in this scenario is given by:

$$R_b = \log_2\left(1 + \Delta \sum_{i=1}^{N} \gamma_i^{(b)} P_i\right)$$

Now, if the channel is being eavesdropped, the transmitter at each DA port uses Wyner's wire-tap coding. It will be logical to assume that Bob and Eve have identical EH modules, such that $\Delta(Bob) = \Delta(Eve)$. Hence the secrecy rate R_s for the user under such an assumption will be:

$$R_s = \left[C\left(\Delta \sum_{i=1}^{N} \Gamma_i^b P_i\right) - C\left(\Delta \sum_{j=1}^{N} \Gamma_j^e P_j\right)\right]^+ \tag{3.5}$$

If we consider a linear energy harvesting model with energy conversion efficiency τ, then energy harvested (in Joules) by Bob for $0 < \tau \le 1$ can be written as:

$$E = \tau(1-\Delta)\sum_{i=1}^{N} \Gamma_i^b P_i \tag{3.6}$$

Now, in a DAS with N-DA ports, the SEE in this scenario will be defined as:

$$\theta_{SEE} = \frac{R_s}{\sum_{i=1}^{N} P_i + P_c} \tag{3.7}$$

In SWIPT systems, in order to ensure that the device successfully harvests the energy from the radio signal, we introduce an energy constraint E_{min} in the optimization problem. Also, θ_{SEE} is a function of power splitting ratio Δ at the receiver module. Hence, the optimization problem is formulated as;

$$\max_{\{P_i\}} \theta_{SEE}$$

$$0 \le P_i \le P_{max,i}, E \ge E_{min}$$

$$0 \le \Delta \le 1$$

Again, as discussed in Section 3.2, to solve this maximization problem, we derive the KKT conditions. For simplicity we assume a fixed value for Δ.

With λ_i, v_i and μ as the parameters, we can define the Lagrangian as;

$$L\left(P_i, \lambda_i, v_i, \mu\right) = \frac{R_s}{\sum_{j=1}^{N} P_{j+} P_c} + \sum_{j=1}^{N} \lambda_j P_j + \sum_{j=1}^{N} v_j \left(P_{max,i} - P_j\right) + \mu \left\{E - E_{min}\right\}$$

Note that, the multiplier μ corresponds to the additional energy constraint $E \ge E_{min}$.

Now, if $\left\{P_i^*, \mu_i^*, \lambda_i^*, v_i^*\right\}$ are the optimal values, then these should satisfy the following set of KKT equations [28];

$$\frac{\partial L}{\partial P_i} = f_i\left(P_1^*, P_2^*, \dots P_N^*\right) + \lambda_i^* - v_i^* + \mu^* \tau \left(1 - \Delta\right) = 0$$

$$\lambda_i^* P_i^* = v_i^* \left(P_{max,i} - P_i^*\right) = 0, i = 1, \dots, N$$

$$\mu^* \left[\tau \left(1 - \Delta\right) \sum_{i=1}^{N} \Gamma_i^b P_i - E_{min}\right] = 0$$

$$P_i^* \ge 0 \text{ and } \left(P_{max,i} - P_i^*\right) \ge 0$$

where:

$$f_i = \frac{\partial L}{\left(\partial P_i\right)} = -\frac{R_s}{\left(P'\right)^2} + \frac{1}{\left(P'\right)} \left\{\frac{\Delta \Gamma_i^b}{C(B)} - \frac{\Delta \Gamma_i^e}{C(E)}\right\} \cdot P' = \sum_{i=1}^{N} P_i + P_c, B = \Delta \sum_{j=1}^{N} \Gamma_j^b P_j, C = \Delta \sum_{j=1}^{N} \Gamma_j^b P_j$$

Further, in SWIPT environment, it is quite possible that the eavesdropper is also a battery-less device and relies on the energy harvested from the received radio signals. In such a case, it is possible to restrain the eavesdropper's operation at the first place by not allowing it to charge. To achieve this, we subject the given problem to one more energy constraint, which limits the amount of energy harvested by the eavesdropper. Since, we have already assumed that $\Delta(Bob) = \Delta(Eve)$, the energy (in Joules) harvested by the eavesdropper is given by:

$$E_e = \tau \left(1 - \Delta\right) \sum_{i=1}^{N} \Gamma_i^e P_i \tag{3.8}$$

With this information, we can redefine the problem as:

$$\max_{\{P_i\}} \theta_{SEE} \quad E \ge E_{min} \text{ and } E_e \le E_{min} \quad 0 \le P_i \le P_{max,i}, 0 \le \Delta \le 1$$

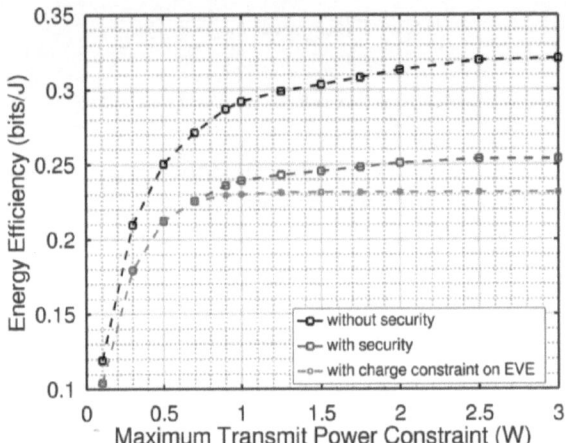

FIGURE 3.7
Energy efficiency with $N = 5$ and $P_c = 1W$ and $E_{min} = 1mW$.

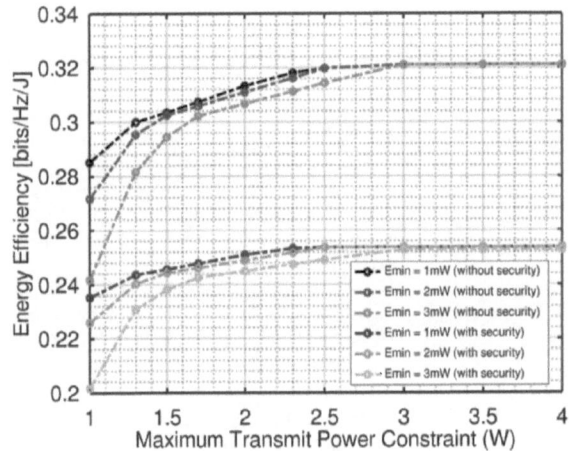

FIGURE 3.8
Energy efficiency w.r.t P_{max} for different values of E_{min}.

In Figure 3.7 energy efficiency for the case of SWIPT is plotted as a function of maximum transmit power constraint on each DA port and it is observed that with charge constraint on Eve energy efficiency further reduces. This is because the transmitter now needs to transmit at slightly lower power levels.

In Figure 3.8, energy efficiency with respect to P_{max} for different values of E_{min} (minimum required energy) is plotted, observe that as the energy required for device charging increases, energy efficiency deteriorates.

3.4 General Case: Multiple Users and Multiple Eavesdroppers

Now, let's consider a general case in which the DAS is serving several IoT devices and an arbitrary number of eavesdroppers exist. If there are a total of K_b legitimate devices in an N DA port DAS, we assume that entire bandwidth is equally segmented into K_b channels and each user communicates over a given channel.

Let's consider two obvious eavesdropper scenarios:

1. In a given channel, only one eavesdropper exists and it does not intercept any other channel.

2. K_e number of eavesdroppers exit in the system, which have the capability to intercept all the existing channels.

In narrow-band IoT standards, FDMA has been implemented for multiple access; hence the achievable rate of device k is given by (Huang, Y., Liu, M. and Liu, Y., 2018);

For case 1:

$$R_k = \frac{1}{K_b} \left[C \left(\Delta_k \sum_{i=1}^{N} \Gamma_{i,k}^b p_{i,k} \right) - C \left(\Delta_k \sum_{i=1}^{N} \Gamma_{i,k}^e p_{i,k} \right) \right]^+ \tag{3.9}$$

For case 2:

$$R_k = \left[\frac{C \left(\Delta_k \sum_{i=1}^{N} \Gamma_{i,k}^b p_{i,k} \right)}{K_b} - \sum_{j=1}^{K_e} C \left(\Delta_j \sum_{i=1}^{N} \Gamma_{i,j}^e P_T \right) \right]^+ \tag{3.10}$$

Where $P_T = \sum_{l=1}^{K_b} p_{i,l}$ and Δ_k, Δ_j are the power splitting for k^{th} Bob and J^{th} Eve respectively.

Again, for simplicity we assume that $\Delta_k = \Delta_j = \Delta$.

Thus, when there are K_b number of devices we can write SEE as:

$$\theta_{SEE,K} = \frac{R_{total}}{P_{total}} = \frac{\sum_{k=1}^{K_b} R_k}{\sum_{k=1}^{K} \sum_{i=1}^{N} p_{i,k} + p_c} \tag{3.11}$$

Now, each IoT device can decode the information only on its own channel but can harvest energy from all the other channels. Hence, the energy harvested by the kth can be written as:

$$E_k = \tau \left(1 - \Delta_k \right) \sum_{i=1}^{N} \Gamma_{i,k} \sum_{j=1}^{K_b} p_{i,j}$$

where: $\sum_{j=1}^{K_b} p_{i,j}$ is the total power consumed at i^{th} DA port and $\Gamma_{i,k} \sum_{j=1}^{K_b} p_{i,j}$ is the power received at k^{th} device from i^{th} DA port. The goal now is to maximize $\theta_{SEE,K}$ with respect to the transmit power at DAS ports and power splitting ratio of the devices subjected to the energy constraint E_{min} for each device and per-DA port P_{max}. This maximization problem is formulated as:

$$\max_{\{P_i\}} \theta_{SEE,K}$$

$$E_k \geq E_{min,k}, 0 \leq \Delta_k \leq 1$$

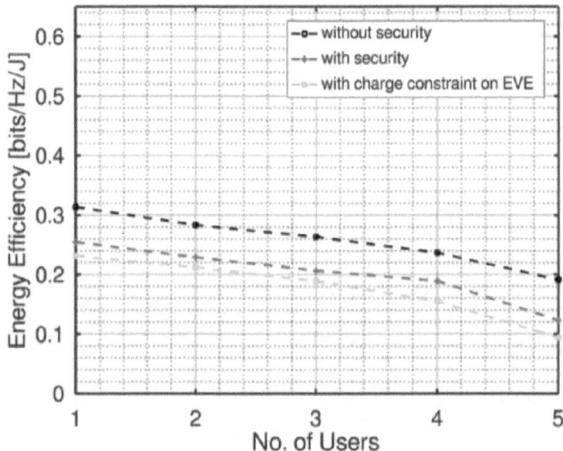

FIGURE 3.9
Energ-Efficiency w.r.t number of users.

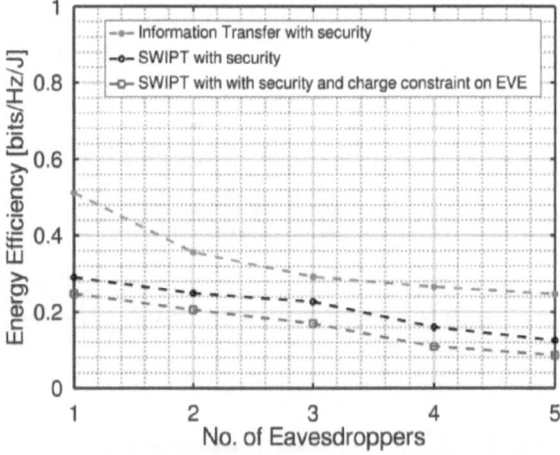

FIGURE 3.10
Energy Efficiency w.r.t number of evesdroppers.

$$\sum_{k=1}^{K} P_{i,k} \leq P_{max,i}, P_{i,k} \geq 0$$

Again, for simplicity, we assume a fixed value for Δ_k. After solving for the optimal solution using SQP, we plot the following results.

From Figure 3.9, it can be observed that energy efficiency deteriorates with an increase in the number of users (K). This is because the achievable rate $R_k \propto \dfrac{1}{K}$ decreases, and transmit power of each DA port increases. Finally, results for the worst-case scenario, where multiple eavesdroppers are considered, are plotted in Figure 3.10. Also, a comparison with the case of Information Transfer is provided.

3.5 Conclusion

In this chapter, we studied energy efficiency and physical layer security aspects in a DAS based IoT network for SWIPT. We used secrecy rate metric of a wire-tap channel model to define SEE. The primary objective was to maximize secrecy rates at minimum possible energy consumption. To achieve this, we formulated SEE as a constrained optimization problem. With the objective function being quasiconcave, we obtained the optimal solution by solving KKT conditions. We considered several different cases, like single IoT device and single eavesdropper, Multiple IoT devices with an eavesdropper in each channel and a worst-case scenario with multiple eavesdroppers with capabilities of intercepting the information channels over the entire bandwidth. Moreover, in SWIPT environment, for an EH eavesdropper an attempt was made to restrain the eavesdropper's operation by not allowing it to charge. In all the cases we observed that there is a trade-off between security and energy efficiency.

References

Ahuja, B., Mishra, D. and Bose, R., 2019. Optimal green hybrid attacks in secure IoT. *IEEE Wireless Communications Letters*.

Bloch, M., Barros, J., Rodrigues, M.R. and McLaughlin, S.W., 2008. Wireless information-theoretic security. *IEEE Transactions on Information Theory*, 54(6), 2515–2534.

Boyd, S., Boyd, S.P. and Vandenberghe, L., 2004. *Convex Optimization*. Cambridge University Press.

Chen, X., Xu, X. and Tao, X., 2012. Energy efficient power allocation in generalized distributed antenna system. *IEEE Communications Letters*, 16(7), 1022–1025.

Choi, W. and Andrews, J.G., 2007. Downlink performance and capacity of distributed antenna systems in a multicell environment. *IEEE Transactions on Wireless Communications*, 6(1), 69–73.

Cover, T.M. and Thomas, J.A., 2012. *Elements of Information Theory*. John Wiley & Sons.

Gopala, P.K., Lai, L. and El Gamal, H., 2008. On the secrecy capacity of fading channels. *IEEE Transactions on Information Theory*, 54(10), 4687–4698.

Guo, S., Wang, F., Yang, Y. and Xiao, B., 2015. Energy-efficient cooperative tfor simultaneous wireless information and power transfer in clustered wireless sensor networks. *IEEE Transactions on Communications*, 63(11), 4405–4417.

He, C., Sheng, B., Zhu, P. and You, X., 2012. Energy efficiency and spectral efficiency tradeoff in downlink distributed antenna systems. *IEEE Wireless Communications Letters*, 1(3), 153–156.

Huang, J., Duan, Q., Xing, C.C. and Wang, H., 2017a. Topology control for building a large-scale and energy-efficient internet of things. *IEEE Wireless Communications*, 24(1), 67–73.

Huang, J., Meng, Y., Gong, X., Liu, Y. and Duan, Q., 2014. A novel deployment scheme for green internet of things. *IEEE Internet of Things Journal*, 1(2), 196–205.

Huang, J., Xing, C.C. and Wang, C., 2017b. Simultaneous wireless information and power transfer: Technologies, applications, and research challenges. *IEEE Communications Magazine*, 55(11), 26–32.

Huang, Y., Liu, M. and Liu, Y., 2018. Energy-efficient SWIPT in IoT distributed antenna systems. *IEEE Internet of Things Journal*, 5(4), 2646–2656.

Jameel, F., Haider, M.A.A. and Butt, A.A., 2017, October. A technical review of simultaneous wireless information and power transfer (SWIPT). In *2017 International Symposium on Recent Advances in Electrical Engineering (RAEE)* (pp. 1–6). IEEE.

Jameel, F., Wyne, S., Junaid Nawaz, S., Ahmed, J. and Cumanan, K., 2018. On the secrecy performance of SWIPT receiver architectures with multiple eavesdroppers. In *Wireless Communications and Mobile Computing 2018*.

Jiang, C. and Cimini, L.J., 2013. Energy-efficient transmission for MIMO interference channels. *IEEE Transactions on Wireless Communications*, 12(6), 2988–2999.

Kim, H., Lee, S.R., Song, C., Lee, K.J. and Lee, I., 2014. Optimal power allocation scheme for energy efficiency maximization in distributed antenna systems. *IEEE Transactions on Communications*, 63(2), 431–440.

Lee, S.R., Moon, S.H., Kim, J.S. and Lee, I., 2011. Capacity analysis of distributed antenna systems in a composite fading channel. *IEEE Transactions on Wireless Communications*, 11(3), 1076–1086.

Leung-Yan-Cheong, S. and Hellman, M., 1978. The Gaussian wire-tap channel. *IEEE Transactions on Information Theory*, 24(4), 451–456.

Mukherjee, A., Fakoorian, S.A.A., Huang, J. and Swindlehurst, A.L., 2014. Principles of physical layer security in multiuser wireless networks: A survey. *IEEE Communications Surveys & Tutorials*, 16(3), 1550–1573.

Ng, D.W.K., Lo, E.S. and Schober, R., 2013. Wireless information and power transfer: Energy efficiency optimization in OFDMA systems. *IEEE Transactions on Wireless Communications*, 12(12), 6352–6370.

Nocedal, J. and Wright, S.J., 2006. Sequential quadratic programming. *Numerical Optimization*, 529–562.

Oliveira, D., Gomes, T. and Pinto, S., 2018, April. Towards a green and secure architecture for reconfigurable IoT end-devices. In *2018 ACM/IEEE 9th International Conference on Cyber-Physical Systems (ICCPS)* (pp. 335–336). IEEE.

Perera, T.D.P., Jayakody, D.N.K., Sharma, S.K., Chatzinotas, S. and Li, J., 2017. Simultaneous wireless information and power transfer (SWIPT): Recent advances and future challenges. *IEEE Communications Surveys & Tutorials*, 20(1), 264–302.

Shannon, C.E., 1948. A mathematical theory of communication. *Bell System Technical Journal*, 27(3), 379–423.

Shannon, C.E., 1949. Communication theory of secrecy systems. *Bell System Technical Journal*, 28(4), 656–715.

Wyner, A.D., 1975. The wire-tap channel. *Bell System Technical Journal*, 54(8), 1355–1387.

Xing, H., Liu, L. and Zhang, R., 2015. Secrecy wireless information and power transfer in fading wire-tap channel. *IEEE Transactions on Vehicular Technology*, 65(1), 180–190.

Yu, H., Guo, S., Yang, Y. and Xiao, B., 2017. Optimal target secrecy rate and power allocation policy for a swipt system over a fading wire-tap channel. *IEEE Systems Journal*, 12(4), 3291–3302.

Zappone, A. and Jorswieck, E., 2015. Energy efficiency in wireless networks via fractional programming theory. *Foundations and Trends® in Communications and Information Theory*, 11(3–4), 185–396.

Zhang, J. and Wang, Y., 2010, May. Energy-efficient uplink transmission in sectorized distributed antenna systems. In *2010 IEEE International Conference on Communications Workshops* (pp. 1–5). IEEE.

Zhao, Y.S. and Chao, H.C., 2018. A green and secure IoT framework for intelligent buildings based on fog computing. *Journal of Internet Technology*, 19(3), 837–843.

4

Blockchain Architecture for Securing the IoT

Rashmi Jain

CONTENTS

4.1 Need of a Decentralized Approach in the IoT

IoT is providing ample opportunities and viable benefits to businesses in current market scenarios. It includes the techniques for collection of data from various devices in various ways. The traditional IoT systems use a centralized mechanism to govern their operation. The working of these systems is broadly divided into four layers, and security of these has to be provided layer-wise.

4.1.1 The Architecture of IoT in Centralized Approach and Security Levels

A typical IoT architecture broadly consists of four components (Suo, 2012). These components can be arranged in four layers for understanding and developing the facilities for IoT. When we talk about the security of the IoT system, we have to think about its security at all four layers:

- *Sensors*
  The sensors or devices are responsible for collecting the data from different devices deployed in the area of interest. As these devices have low energy and limited computing power, it is difficult to apply a powerful security algorithm to protect their data from attacks like DoS. So, a simple, less computing-intense requirement encryption algorithm can be deployed here along with an agreed key between the sender and the receiver. Hop-By-Hop encryption can be applied to protect sensor data.
- *Internet*
  A network that is needed to communicate the data collected at a central place is the next layer. DDoS and Man-in-the-Middle are the commonly seen attacks at this level, which can be secured by Anti-DDoS, identity authentication, encryption, and other communication security mechanisms.
- *Server*
  It stores and manages the data communicated by sensors over the network. Usually, cloud servers are preferred for this purpose. The server analyzes this data and makes decisions for appropriate action. It has to identify malicious data and take care so that it will not harm the system. Secure cloud computing algorithms and anti-virus software can be used to provide security at this layer.
- *Application program*
  Application program accesses the resultant data with the server for further utility. Different application programs are designed to access the data in different ways with suitable authentication algorithms, key management methods and privacy protection techniques.

Currently, IoT systems use centralized access control systems popularly control their functioning. Some devices are generally within the same trust domain where their human operators are located, and these can be controlled finely with our traditional client–server system.

4.1.2 Decentralized Approach to Address Issues in Centralized Approach

IoT systems generate huge volumes of data, which is collected by various sensors from the field. Communication of these data involves hundreds of billions of transactions with the

centralized data servers. The cost associated with this communication, and the installation and maintenance of a suitable sever, is high. A centralized issues approach is stated below (Gokhan, 2018):

- Some applications are more dynamic than the traditional systems (i.e., the devices are mobile, may belong to different owners or organizations during their tenure of work, or the devices may be controlled by multiple controllers simultaneously). In such scenarios, this centralized access control system may not seem promising from performance and security perspectives.
- Centralized access control also incurs more expenditure for maintenance of the cloud server. Data generated by sensors are voluminous, which adds yet more cost to storage and processing.
- If the IoT application is time-critical, the cloud server is often not able to support it. Even when the general cost and technical challenges are taken care of, cloud servers will pose the problem of access delays.
- If these centralized servers fail, they will disrupt the entire network. SPoF issues prevail in such a case.

A decentralized blockchain-based approach to IoT networking addresses many of the issues mentioned. Blockchain technology adopts a standardized peer-to-peer communication model. This technique distributes computation and storage needs across a large number of devices in the network. Thus, the network can be saved from a SPoF. The system may consist of different administrative domains that control the IoT network according to their needs. The network architect defines the access control mechanism for the network, and distributes this access control information to different devices in the network. Users have to follow the rules while using the facilities. Different entities working in consensus will allow an entity to access any of the network services. Smart contracts govern the functionality of the system, simplify the whole process, and reduce the communication overhead between the nodes. As the current state data is available to all the nodes, they can exchange information in real-time and reduce the delays.

4.2 Blockchain Technology

Blockchain was first introduced in 2009 by Satoshi Nakamoto (Novo, 2018). Bitcoin, a cryptocurrency (electronic cash), was the first and most popularly used implementation of this technology. For a successful decentralized approach, it must support the following fundamental operations.

4.2.1 Fundamental Operations

Blockchain Technology has the potential of replacing the currently used technologies in industries and facilitating new business structures. So, the strategists, planners, and decision-makers across industries and businesses should focus on the ways of applying this technology to different applications. The concepts on which the blockchain technology works and provides a way of committing transactions—that it will be more secure due to peer-to-peer transfer, transparent to other users, highly tamper-proof, auditable, and efficient—are discussed below.

- *Peer-to-Peer Messaging:* All the nodes have dedicated links to share the data between them. Data transfer is secure over a dedicated link also the delay is less.
- *Distributed File Sharing:* The current and all previous status of transactions are available with all the nodes. So, in case, any device fails, it will not affect the operation of the rest of the network.
- *Autonomous Device Coordination:* Special programs called Smart Contracts are designed to streamline communication among devices. By using these smart contracts and consensus algorithms, device coordination is done.

4.2.2 Working of Blockchain

A blockchain contains a sequence of blocks in which each 'block' is numbered and includes information about the new operations to be included in the block and where the hash of the block is calculated, by using a secure hash function, and the previous block's hash. The chain of blocks is created from the first block—the 'genesis' block—to the current block. The genesis block is generally hardcoded into the software. Figure 4.1 below shows the structure of a simple blockchain. Blocks contain a set of transactions that denotes the operations to be performed or values to be transferred to other entities. The transactions from different entities are sent to the network and gathered into a block. After verification of data in a block, it is added to the existing chain. All transactions are visible to all the peers in the blockchain, and everyone has precisely the same copy of information available in the chain. Pool miners or solo miners mine validate the transactions included in a block. Mining is a resource-intensive and complex task in the blockchain. The complexity of the mining process will decide how resistant the system is to any kind of tampering. Every block contains a challenge-response based Proof of Work (PoW) (Novo, 2018). A PoW in a block is a complex mathematical operation needed to validate the block in the blockchain. The PoW generated by a node is verified by the rest of the miners when they receive a block. Mining ensures that the node added to the chain is secure, tamper-resistant and attains consensus. Miners charge some fee or some coins for the validation work done. The consensus is a crucial feature in decentralized systems which require that two or more nodes mutually reach an agreement on a proposed value needed for computational

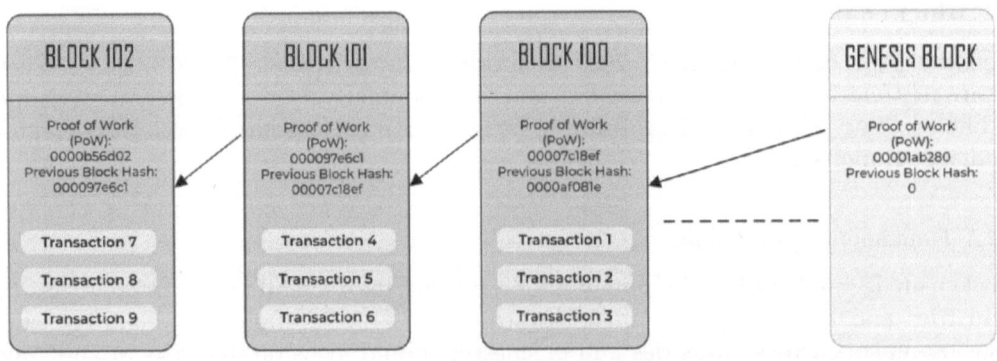

FIGURE 4.1
Blockchain structure.

purposes. Among various consensus methods like PoW, Proof of Stake (PoS), Byzantine Fault Tolerance (BFT), Practical Byzantine Fault Tolerance (PBFT), Binary Consensus etc., PBFT and Binary consensus methods are preferable in IoT applications (Zoican et al., 2018). The amount of computation should be smaller in IoT based systems as they are resource-constrained. These protocols guarantee that for every block, a specific level of difficulty is attained to generate a new block. PoW is a computing-hungry process which may not be suitable for IoT, so IoT applications use a simplified version of this: Simple PoW.

4.2.3 Features of Blockchain

Blockchain offers some features that are beneficial for IoT application. And therefore, it can be used to develop secure IoT systems. Those features are as given below:

- *Decentralized Approach*
  Not a single node or entity gains the control of the complete system. All the participating entities take part in controlling the operation of the system.
- *Distributed and Transparent Data*
  The current status of data in a blockchain is available with all the nodes. The blocks store the set of all transactions committed till the recent one and nodes distribute all information publicly among the peers.
- *Auditable Transactions*
  Every node can verify the operations performed by all other nodes. Each user has to digitally sign the transactions to be included in a block. Digital sign verifies who performed each transaction. The current block has a hash value computed over the hash of previous block.
- *Consensus Algorithms*
  Every node participates in a consensus process to validate a transaction. If the majority of nodes agree for a specific operation, then only that operation can be performed. A single central entity does not control the transactions to be completed. The preferred consensus algorithms for IoT application are BFT, PBFT or Binary Consensus algorithm.
- *Security*
  It is very challenging to make any changes in the transactions done previously, as all entities are party to them. Digital signatures are put with all the transactions done prior, and the hash value generated in the current block is computed based on the previous block's hash, along with current block transactions. Therefore, tampering with the data is very challenging for any malicious user.

4.2.4 Types of Blockchain

Two major types of blockchains can be used, depending upon the nature of the application, whether public or private (Novo, 2018). Hybrid blockchain combines the features of public and private blockchains according to the need of the application.

- *Public Blockchain*
  In this type, every node on the chain can read or modify the data. There is no requirement of any identity of the node available in the blockchain. Different applications may provide simply read permission to all nodes, and some special nodes get read as well as write permission. The most successful Bitcoin application provides write as

well as read permissions to all the nodes, and is the best example of a public block-chain. After Bitcoin, several other cryptocurrencies were designed and are in use, such as Litecoin, Omni, and Gridcoin.

- *Private Blockchain*
  This type of structure can be deployed in an organization where all the users are trusted and known to each other. Here the rights of users may vary according to the hierarchy to identify the roles of participants. This type of blockchain can be used in asset movements in financial marketplaces or in supply chain applications, and so on.

- *Hybrid Blockchain*
  This type of blockchain has security and transparency features of public blockchain along with data privacy feature of private blockchains. Applications can decide which data to share publicly and which to keep secret. A ledger of a hybrid block-chain is available to all users, but access control rights are not given to all of them. XinFin, for the management of supply chain logistics, Dragonchain, Ripple network and the XRP token are some hybrid blockchain examples.

4.3 Application of Blockchain to IoT

Blockchain technology can address the current issues of IoT like security and reliability. In small-scale IoT applications, peer nodes can represent different devices and can be connected as nodes in blockchain architecture. Transaction processing is enabled in coordination with connected devices, and all peers can have the updated information. With a decentralized approach, (SPoF) is eliminated. A robust ecosystem for IoT is supported for the devices to run on. The cryptographic hash functions used by blockchain would make consumer data more secure (Banafa , 2016). Transactions would be immutable and transparent to all users of the system. The ledger of IoT in blockchain can take a form, as shown in Figure 4.2 below. The ledger is tamper-proof and cannot be manipulated by malicious

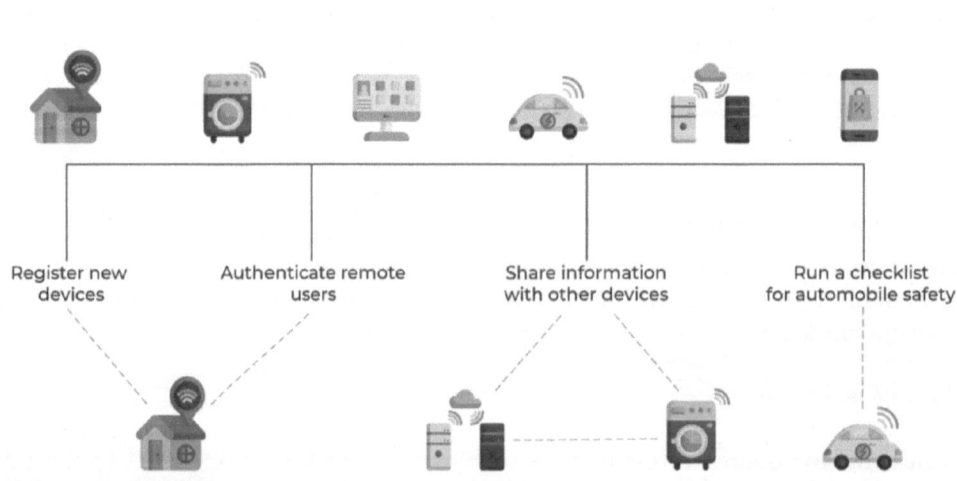

FIGURE 4.2
Transactions in blockchain for IoT.

actors because it doesn't exist in any single location, and Man-in-the-Middle attacks cannot be staged against it because there is no single thread of communication that can be intercepted. But this ledger size may grow huge when the number of devices connected to network increases (Banafa, 2016). In the long run, the size of the database available with the devices would consume an ample storage space, and it may not be possible for resource-constrained devices to provide for it. For addressing this issue, another approach can be used where sensors or low-layer devices are not directly connected to the blockchain. Instead, use some mediator machines or devices.

4.4 Blockchain-Based Architecture for IoT Network

As the end devices, collecting data from the field sensors have low power and low computing capability, they cannot be directly connected to the blockchain network as peer entities. The peer entities have to store massive information about all the transactions done till then, and validate the transactions proposed by other entities. These tasks need significant storage and computing power. As a consequence, the low power sensing elements will not be able to perform the functions intended for a peer entity. For this reason, the sensing devices are connected to different intermediate nodes (Novo, 2018). Figure 4.3 shows the components and their interconnection with each other in the network. It consists of the following parts:

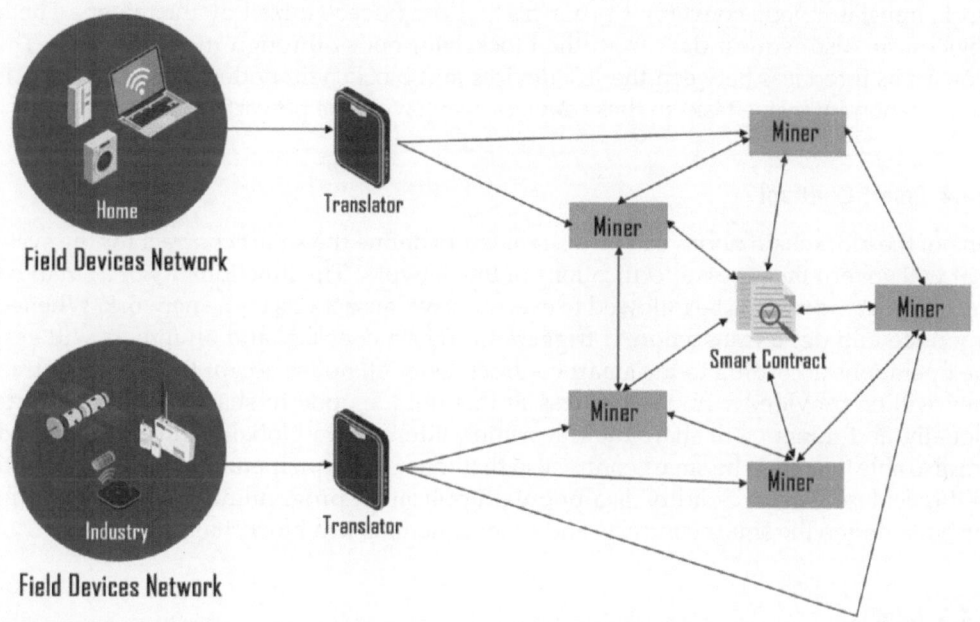

FIGURE 4.3
Blockchain architecture for IoT.

4.4.1 Field Devices

The actual sensing IoT devices gathering data from the field in which they are deployed, are the field devices. As discussed earlier, they are low power devices and do not get connected to the blockchain network directly. Instead, they register themselves with some lightweight intermediate nodes that convey the field data to other nodes in blockchain network. There is no need for a third-party application to exchange the data among devices. For communication of this data to intermediate nodes secure IoT communication protocol such as Constrained Application Protocol (CoAP) can be used (Jain and Jain 2019).

4.4.2 Intermediate (Mediator) Nodes

These are lightweight nodes that collect data from the field devices registered with them. All field devices must get registered with some intermediate node so that they can send data to nodes in blockchain. A field device may get registered with more than one mediator node to share its data. Once the device gets registered with a mediator node, it decides the access control mechanism for communicating with different field devices. It takes the help of translator nodes to analyze the data received from field nodes and converts it to a form that would be understood by the miners in the network. Being lightweight, these nodes again do not store the data of blockchain and do not participate in the validation of a transaction like miners. Any other entity can also work as a mediator node (miner or translator). The usage of the mediator node helps in providing scalability to the IoT system.

4.4.3 Translator Interfaces

The data received from the field devices may use CoAP or some other protocol to transmit the data securely to mediators. To forward this encrypted data on to the miners in the network, translator node converts it to a form that can be recognized by the miners. The IoT devices can also request data from the blockchain nodes through the translators. Thus, they act as interfaces between the IoT devices and blockchain nodes. The translation is a computation-intensive task, so these devices also have to be powerful.

4.4.4 Smart Contract

One of the blockchain nodes can be designated to define the smart contract for the system that will govern the access specifications of the network. The functionality of a smart contract is to set the rules to be followed to execute any transactions in the network. Whenever any node initiates a transaction, it triggers the smart contract, and all miners will verify the operations according to the smart contract. Once all nodes accept the smart contracts, they will be provided with the address of this unique node to share the smart contract globally, and miners will share the transaction information globally. Intermediate nodes are also able to access the smart contract so that they can implement related functionalities in the field networks. 'Solidity' is a popular application programming environment preferred to define the smart contracts and other functions in a blockchain network.

4.4.5 Miners

These nodes should gather transactions initiated by different nodes that are not committed in a block. These are the significant nodes that validate all the transactions initiated by any node in the network. They use the smart contract for validating the operations and are responsible for maintaining or sharing the blocks globally. The confirming process depends on the consensus mechanism used. BFT is a suitable protocol for IoT applications, so it is used for deciding whether a particular transaction should be included in a block or not. As all the miners share the same copy of transactions performed till then, it is difficult to tamper with them, and the data remain secure. Thus, the miners feature the properties of the blockchain network like decentralized information, security, tamper-resistant, etc. Above mentioned components can be used to develop a small scale IoT system based on blockchain technology.

This type of structure of IoT may contain a large number of sensing devices deployed, that may generate a massive amount of data regularly. Keeping complete data with different nodes in a block will increase the size of the block, and downloading this at all nodes will become an intensive task for them. The amount of data generated will increase to such an extent that the nodes will not be able to handle the pace with the limited capacity of the nodes. In the most successful implementation of blockchain technology, Bitcoin, the size of the chain at the time of writing is around 285 gigabytes, and it is still growing. Downloading this much data is tedious for the nodes. To deal with this fact, some alternative arrangement can be utilized as mentioned in the next section.

4.5 Blockchain-Based Architecture of IoT Network with Decentralized Cloud

In Industrial IoT systems where IoT devices are used on a large scale, the data is generated by them is enormous. According to a study by Gang Wang (Gang, 2019), near about 302.4 GB of data is generated by a medium-sized IoT network per week, whereas in the most popular Bitcoin application, on average, 1GB data is produced in a week. So, storing this much data in IoT blockchain nodes will not be a feasible solution. Using the storage capacity of a cloud can be an excellent remedy for this. To avoid a SPoF, decentralized clouds can be deployed in such applications. Figure 4.4 shows this type of architecture. It divides the network into three broad sections.

4.5.1 Field Network

The first section is local IoT network that consists of the field devices and the intermediate nodes, as stated earlier. This network of sensors and low-end devices is connected to intermediate nodes to sense the data from the field. The translator node communicates this data to a miner node of the blockchain, which is further connected to such other nodes in the chain.

4.5.2 The Blockchain Network

The second section, blockchain network, comprises mining nodes and the designated nodes for maintaining the smart contract and shares the global status of the blockchain among all the peers on the network. The translator nodes connect the IoT network to the blockchain network. The nodes in this section store the recent blocks in the blockchain that are needed to validate the upcoming blocks. These nodes can move the older chain of blocks to clouds. The third section contains Clouds and Cloud Interface Nodes as its

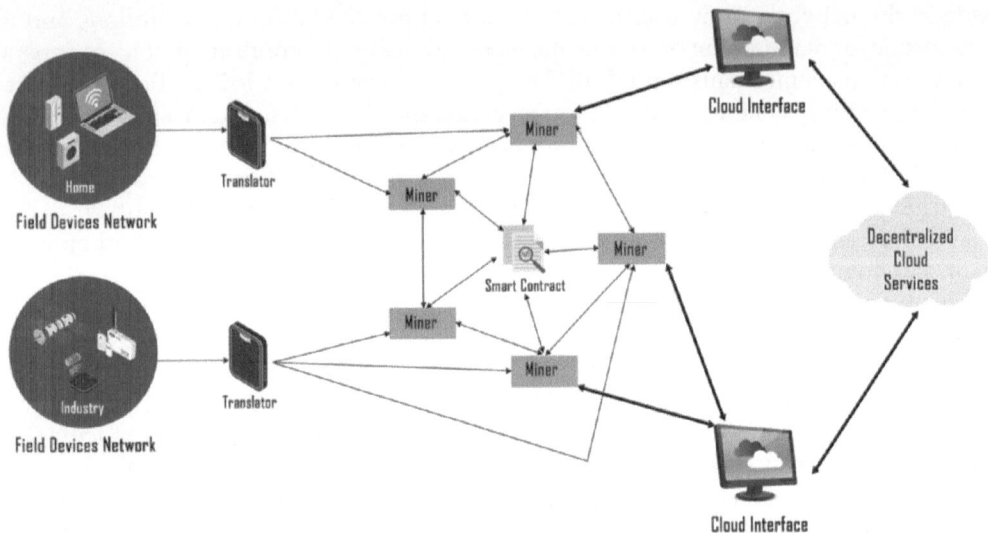

FIGURE 4.4
Components of blockchain-based IoT architecture with cloud infrastructure.

components (Gang, 2019). The blockchain nodes are connected to clouds via cloud interfaces to get the benefit of clouds' unlimited storage. The cloud infrastructure is also decentralized, and a decentralized object storage system is needed to store and access the data over this cloud infrastructure. The blockchain network serves as the bridge that connects local IoT networks and the clouds. Functions of cloud and cloud interfaces are as follows:

4.5.3 Decentralized Cloud

The third section has decentralized cloud and cloud interfaces as its components. Decentralized cloud provides virtually unlimited processing and storage capacity to the blockchain nodes. Using an interface, the translator and the mediator nodes access cloud facilities through the blockchain network. Cloud provides different services like data storage, data processing, management of data as and when needed by blockchain nodes. It has to provide suitable Application Programming Interfaces (APIs) to access cloud services. The cloud is managed in a decentralized way by different operators. Along with the specialized servers, the miner nodes or the smart contract managing nodes and the cloud interface nodes can also share their hardware for the creation of the decentralized cloud infrastructure. Storj Network is a well-established system that encrypts shards and distributes data to get stored in a decentralized way (Gang, 2019). It guarantees the consistency of data on the decentralized cloud. BFT is used here for consensus purpose.

4.5.4 Cloud Interface

Cloud Interface is used to synchronize data sharing between blockchain nodes and cloud. It can handle operations such as how a blockchain node requests the services of cloud, how services are provided to the requester, and to check whether the requester is a legitimate user or not. If some error occurred during data sharing, this interface would handle it. In this type of architecture, the system may decide to keep a certain amount of recent blocks' information with the blockchain nodes, depending upon the storage capacity of the machines used as a miner, translator or mediator node. The rest of the information in the blockchain can be preserved on the cloud so that the nodes need not store the complete chain with them, and they can utilize their power for streamlined operation of the blockchain. Multiple users manage the decentralized cloud data by using a decentralized object storage system.

4.6 Benefits of Blockchain to IoT

Though the limited capability sensing devices are not directly connected to a blockchain, this technology can, using the above described architectures, be applied to IoT applications effectively. Different models can be developed for different applications with specifically defined features. Major benefits of applying blockchain technology to IoT systems are as follows:

4.6.1 Fraud Proof

Blockchain allows the tracing of the measurement of autonomous and smart IoT devices. It prevents forging or modification of data as the data is shared among a large number of users on the chain. All the transactions in the ledger are signed and time-stamped. These systems are thus protected from cybercrimes and frauds.

4.6.2 Low Cost

It is possible to exchange data, or execute financial transactions, without the need of a third party or a centralized broker. This helps in reducing the infrastructure cost. Blockchain also reduces the deployment and the operation cost of IoT applications significantly.

4.6.3 Decentralized Control

Decentralization of an IoT infrastructure reduces the amount of data transferred to the central authority for processing, analysis and any kind of settlements. This makes the operations to speed up. Again, all the transactions performed till then are available with each participating node. Therefore, they are not dependent on any other node for any of committed transactions in the ledger.

4.6.4 Security

Security and privacy of the managed data are improved by using trustless messaging between devices in an IoT network. Peer-to-peer communication also supports the privacy of the data shared.

4.6.5 Transparent Data

As all the nodes have the same copy of the blocks and transactions, so failure of any node does not significantly affect the performance of the network. This supports the elimination of a SPoF within the IoT system.

4.6.6 Device Autonomy

Blockchain enables device autonomy by incorporating smart contract, individual identity, the integrity of data, and supports peer-to-peer communication by removing technical bottlenecks and inadequacies.

4.6.7 Distributed Ledger

The capability of having a distributed, trusted ledger of all transactions taking place in the network can be useful for imposing the constraints, regulatory requirements and compliances for a smooth operation of industrial application.

4.7 Challenges to Application of Blockchain to IoT

As blockchain is a comparatively new and yet to be explored technology, there are many challenges while applying it to IoT applications. Some major issues are listed below:

- While integrating blockchain technology with IoT applications, storage issues may arise. An IoT device can generate a huge volume of data in a short period. It requires the storage of both the data hash and the data itself. Over a long period, this blockchain grows, and all the participating nodes need larger storage capacity and higher bandwidth. Maintaining transaction processing and sufficient storage for the ledgers becomes very costly.
- Though scalability can be provided by using mediator or translator nodes, it may be a challenge when the number of resource-constrained devices is massive. Adding new devices is a recurring phenomenon in any IoT application, and the optimal blockchain architecture has to scale to adopt new IoT devices.
- The rate of generation of data is very high in IoT as compared to other blockchain applications, so the rate of adding new blocks should be less.
- Resource constraints of the IoT devices are again a bottleneck for processing-intensive computations of generating data hash and processing the transactions.
- The field devices are not part of the blockchain network directly. Therefore, blockchain technology does not address physical security of these field devices. So, some extra measure and additional care has to be taken.

4.8 A Case Study: Brooklyn Microgrid (BMG)

Many projects have been developed using blockchain technology in IoT systems. Brooklyn Microgrid (BMG) (Orsini , 2020) is one of the best examples of it. Scientists say that,

> "This project..., is the first version of a new kind of energy market, operated by consumers, which will change the way we generate and consume electricity."

BMG is an **energy marketplace** for locally generated, renewable energy. Scientists have developed this model to share the energy generated by community people with community people.

Exergy, a permissioned data platform, was created using some innovative solutions along with the power of blockchain technology. It can send energy across the existing grid infrastructure, which is generated by Brooklyn residents. People participate in a simulated energy marketplace where people are willing to pay for locally generated, renewable energy. The solar grid, producing electricity, can be monitored with a set of IoT system, which will measure the electricity generated, share electricity with other users, and share the burden of maintenance. Exergy builds a reputation through its history of records and exchanges. People are producers as well as consumers and they care for their community's energy future.

4.8.1 The Exergy Network

Exergy platform has the potential to inspire the energy model of the future, and already the possibilities seem endless. It is the network of prosumers and consumers, and there is no conventional central plant to produce and share energy as a utility. The entities on the Exergy network are as given below:

- *Prosumers*
  On the Exergy platform, prosumers are generating energy with their renewable resources. They can share power autonomously with consumers in near-real-time on the platform in their local marketplace. Local businesses and residents can be the prosumers. They take care of devices, ensure that devices are recording actual solar production, and share the burden of maintenance.

- *Microgrid*
  A microgrid is an ecosystem of connected prosumer and consumer energy assets. Energy is generated, stored, and transacted locally, creating more efficient, resilient, and sustainable communities.

- *Distributed System Operator*
  The distributed system operator has the task of managing energy usage, fulfilling the demands at negotiated rates, and load balancing. It has the access granted to consumer data available in the management systems.

- *Electric Vehicle Charging*
  When a charging station, either public or private, or even an electric vehicle, has a surplus of energy, it can share it and make it available for purchase on the local network. Mobile apps can be used to send alerts to consumers according to the budgets set by them.

- *Consumers*
  Residential users or local business organization can be the consumers. They can access the energy at a negotiated rate from various prosumers. A messaging system can be used to inform the user about the availability of energy and the requirement of the user.

4.8.2 Working of Microgrid

Figure 4.5 shows the connection of the entities of the microgrid. Prosumers generate the energy and connect it with smart meters that are used to measure the amount of energy

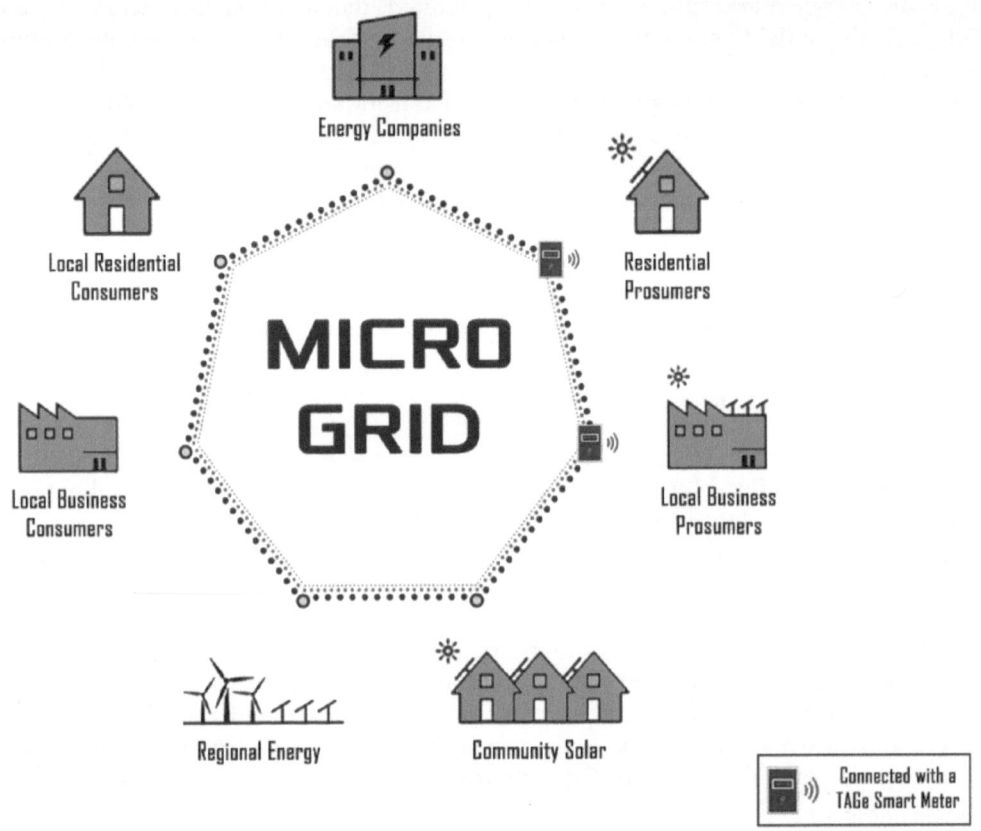

FIGURE 4.5
Entities on a Brooklyn microgrid.

transacted with the consumers. The community generates solar energy. Smart contracts implemented govern the sharing energy between prosumers and consumers. Rate contracts are used for selling of extra energy generated, which is cheaper than the energy supplied by the conventional electricity suppliers. Regional energy companies and other private companies can also access the energy generated by the prosumers if they have extra energy available with prosumers.

A consumer can approach a prosumer for an amount of electricity needed at a particular time, over the peer-to-peer network, and in response, the prosumer will supply that electricity. A new block will include the transactions to be performed. This block is tamper-proof and trusted. Thus, this project provides benefits in environmental, cost, and reliability terms.

Because of their ability to offer consistent, secure, and high-quality energy supply, microgrids have been receiving much attention recently. They are also providing sustainability and energy efficiency. They are also able to take advantage of shared energy storage systems in the community to improve operational reliability and economy. The cryptocurrency model is efficiently deployed to develop a platform for trading energy.

4.9 Conclusion

With the frequent use of IoT in domestic, industrial applications, the security of data communicated over the network is crucial. Among various techniques of securing the data in IoT, Blockchain is a promising solution. Blockchain technology has drawn a lot of attention from developers from the industry and the academia for using it in IoT systems. Here, instead of maintaining a huge volume of data in centralized servers and clouds, this decentralized approach is used, which is capable of addressing the scalability, privacy, and reliability concerns in the IoT. Systems such Brooklyn Microgrid, utilizing the concept of a decentralized approach supported by blockchain are proving to be innovative methods of generating and distributing energy. They can be efficiently adopted in future for other applications. However, the problem of storing and managing the blockchain in IoT networks persists on account of the massive amount of data generated from IoT applications, and the limited resources in the IoT infrastructure. Decentralized cloud services and partial storage of blockchain data in its nodes could address this issue.

References

Banafa Ahmed, 2016, "How to Secure The Internet of Things with Blockchain", *Datafloq Newsletter*, Hague, The Netherlands, Available at: https://datafloq.com/read/securing-internet-of-things-iot-with-blockchain/2228

Orsini Lawrence, 2020, "Brooklyn Microgrid Project", Brooklyn, NY 11217, Available at: https://www.brooklyn.energy/

Gang Wang, Zhijie Jerry Shi, Mark Nixon, Song Han, 2019, "ChainSplitter: Towards Blockchain-based Industrial IoT Architecture for Supporting Hierarchical Storage," *Proceeding of 2019 IEEE International Conference on Blockchain*.

Gokhan Sagirlar, Barbara Carminati, Elena Ferrari, John D. Sheehan, Emanuele Ragnoli, 2018, "Hybrid-IoT: Hybrid Blockchain Architecture for Internet of Things – PoW Sub-blockchains," *IEEE Blockchain International Conference*

Suo Hui, Jiafu Wan, Caifeng Zou, Jianqi Liu, 2012, "Security in the Internet of Things: A Review," *International Conference on Computer Science and Electronics Engineering*.

Novo Oscar, March 2018, "Blockchain Meets IoT: an Architecture for Scalable Access Management in IoT", *Journal of Internet of Things, Class Files*, 14(8) pp. 1184-1195, doi: 10.1109/JIOT.2018.2812239.

Jain Rashmi, Jain Shweta, Nov. 2019, "A Survey on Security Mechanism to Embedded Platform Based IoT Devices", *Proceeding of International Conference on Innovation in Engineering Sciences, Management and Technology*, Nagpur, India.

Zoican Sorin, Vochin Marius, Zoican Roxana, Galatchi Dan, Nov. 2018, "Blockchain and Consensus Algorithms in Internet of Things", *International Symposium on Electronics and Telecommunications*, Timisoara, Romania, 978-1-5386-5925-0/18, IEEE.

5

Reliable Data Transmission Using Biometric Authentication Methodology in IoT

Ambika N

CONTENTS

5.1 Introduction

Internet-of-things (Atzori, Iera, & Morabito, 2010; Alaba, 2017) is a trending technology that aids in providing a common platform for different kinds of devices. The devices will be able to communicate using this platform. This technology supports many applications including industry (Da Xu, He, & Li, 2014), health monitoring systems (Nguyen, Mirza, Naeem, & Nguyen, 2017), parking (Khanna & Anand, 2016), public transport (Melis, 2016), and so on.

Health monitoring systems is a critical area. The health of individuals is an essential part of society. Various authors have offered many suggestions to improve the health monitoring systems (Arcelus, Jones, Goubran, & Knoefel, 2007) available. The recommendations also aim to reduce human effort. Some of the essentials of the health monitoring system are

- *Security to data* – Security (Andrea, 2015) has become prior importance to protect the data from different kinds of attacks launched by intruders. The data stored has to be confidential. The data should not be available to anyone other than the respective authenticators.

- *Reliability* – The practicing professionals have to be more accurate in analyzing the symptoms and providing appropriate medication. In the era of computers, the hospitals store the data in digital form for convenience. Failing to protect these data can lead to inaccuracies. Such a situation could prove fatal for the patient, so better measures have to be considered to protect the data. The preliminary test that can be adopted is authentication.

Authentication can be one-way or mutual. The procedure is a process where the communicating party is confirming its identity to the other party during this procedure. In one-way authentication, only the client will be verifying its identity, whereas in mutual authentication procedure, both the parties affirm their identities before commencing with the transmission of data.

The proposed work uses biometrics evaluation (Nelson, 2013). The procedure begins with the extraction of biometrics from staff and the patient. The hash value is generated using the sample collected. The server stores the generated hash values. Whenever the staff member examines the patient, he authenticates himself by giving the biometrics. The hash value for the same is generated and cross-verified with the data stored on the server. On positive confirmation, personnel are allowed to update the evaluation summary of the patient. The same is updated using the hash value of the biometric extracted and location information of the patient. This value is concatenated with the data and submitted to the server. The methodology aims to authenticate and increase the reliability of the system.

The proposed work evaluates the health monitoring system. The suggestion aims in improving safety and security to data by transmitting the hash values of the personnel and the location for authentication. The work explanation falls into four sections. A literature survey follows introduction in part 2. Part 3 details the simulation of the work. Part 4 is the conclusion.

5.2 Literature Survey

Many authors have suggested using different methodologies to authenticate the communicating parties before commencing the transmission of data. The article provides a detailed description of the same.

Nyberg's one-way accumulator (de Meer, Liedel, Pöhls, Posegga, & Samelin, 2012) aids in securing the devices (Yao, Han, Du, & Zhou, 2013). The version is a one-way hash function that has a quasi-commutative property (Lian, 2009). The output is obtained by performing bit operations on them. The system stores the participants and the hash values. The hash values aid in authentication.

The suggestion consisted of a decentralized anonymous procedure (Alcaide, Palomar, Montero-Castillo, & Ribagorda, 2013). The clients of the system co-operate to generate the private key. This is followed by RSA pair key generation. The key generated will not get compromised as the generation is the collaborative effort of all the clients. The clients of the system can gain access to an anonymous set of certified attributes. The central organization is responsible for distributing the certified characteristics. These characteristics verify themselves with data collector entities, and the servers can verify the communicating parties using attribute-based Boolean formulas. The servers provide the flexibility to change these attributes according to their convenience.

The suggestion uses identity authentication (Mahalle, Anggorojati, Prasad, & Prasad, 2013). The system handles access control in the work, and uses WiFi facility (Balasubramanian, Mahajan, & Venkataramani, 2010). The work uses one-way key authentication. Elliptic curve cryptography and Diffie–Hellman algorithm (Hankerson & Menezes, 2011; Kumar & Singh, 2016) is used in the generation of a secret key. The work facilitates two communicating devices' authentication, and a token is provided by the system to the devices to provide access control.

The network uses XOR manipulation (Lee, Lin, & Huang, 2014). In the proposal, the devices have embedded RFID technology connected to the internet, and the reader then evaluates the identity of the device using an authentication procedure.

An implicit certificate-based authentication methodology proposes to secure IoT applications (Porambage, Schmitt, Kumar, Gurtov, & Ylianttila, 2014b). The proposal supports heterogeneity and scalability. The scheme runs in two phases, the first of which is known as the registration phase. This phase collects the security credentials from the trusted entities, and the requestor's identification acceptance follows a certificate issue. The second phase is the authentication phase, when the procedure validates the communicating parties. The client communicates with the secret writing suites and their identity. The output is cross-verified against the records stored in the server. On successful completion, the client sends its certificate along with random cryptographic nonce and the MAC value. The public key calculation follows the positive verification of the MAC attachment. The server calculates the standard key using the client's public key and its private key. The server transmits the certificate attached to the nonce and MAC attachment.

The suggestion is a user-friendly approach (Petrov, Edelev, Komar, & Koucheryavy, 2014). The near field communication technology tags store the essential credentials. The device consists of a cluster of encrypted passphrases. These entities provide a set of rights accessible by them. The work uses a robust symmetric algorithm. The certificate center is the control point in the work. The algorithm runs in seven phases. The setup phase encompasses the public key generation. The center is responsible for generating the master key, and also creates secret keys for the clients and departments. The system uses a sizable unused prime number to add a new department, and the secret key generation uses the output of the previous stage. The same procedure aids in generation of user addition. A random passphrase is created by the department to provide access to the user. The user authenticates itself using the user ID and critical credentials. If the user is not a trusted entity, the respective department has the power to nullify its right and detach the it.

The proposal suggestion aims in establishing secure links between the communication devices (Porambage, Schmitt, & Kumar, 2014a). Its foundation lies in implicit certificates. The procedure consists of two stages: during the registration stage, the cluster members obtain the certificates from the respective group head and generate public/private keys, and then the second phase is authentication. The proposal includes three scenarios. Using either of the verification scenarios, the communication is performed followed by transmission of data.

The mutual authentication procedure aims to provide security (Kalra & Sood, 2015a). The validation happens between embedded devices and servers using the HTTP protocol. The proposal based on elliptic curve cryptography has three phases. The procedure commences with the registration phase, where the embedded device gets itself enrolled with the cloud server. In this procedure, the server stores its cookies on the device. The second phase is known as the pre-computation and login phase. During this phase, the device tries to connect itself with the server and transmits a request to the respective server. The third phase is authentication, and during this, the two communicating parties are mutually verified for their identity by using elliptic curve cryptography parameters.

The proposal (Moosavi, Gia, Rahmani, Nigussie, Virtanen, Isoaho, Tenhunen, 2015) aims in better health monitoring. The system aids in monitoring using e-health gateways, in which the patient will carry an inbuilt sensor that picks up and processes many variables. A precise judgment conclusion is made after the unusual condition of patient identification, and a medical sensor network installed in the environment collects the readings and transmits them to the available gateway. The gateway is a mediator between the network installed in the hospital and the internet, and performs functions including grouping of data, filtration, and reduction. The back-end system comprises cloud computing platform, warehouses, and data analytics. The system synchronizes constantly over time. Data is classified as public and private. The clients enable web applications to access the corresponding data, and the proposal suggested then has multiple roles for the gateways. They communicate using different wireless protocols to accomplish the task, and tasks assigned to them include acting as a local repository, storing temporary data, enabling intelligence.

The work discusses sensor-based communications and considers the third-trusted authority(Hou & Yeh, 2015). The user registers once and their set of certified parameters is then preserved. The login coupon uses the generated parameter. Using this coupon system resources are assigned, and the reclaim of services of the server is a part of the work. The coupon acts as a single sign-on (SSO) availability. Usage of tokens aims to access multiple services, and the work adopts one-way hash function and choosing a random nonce. This facility adopts efficiency and strong security for the system.

An architectural model suggestion aims to secure the system (Barreto, Celesti, Villari, Fazio, & Puliafito, 2015). The work considers two scenarios: accessing the IoT devices directly, and accessing the same through cloud services (Aazam, Khan, Alsaffar, & Huh, 2014). The proposal considers the identity-provider–service-provider model. The system provides a single sign-in option, and using this authentication methodology, the system can gain access to all the service providers that are accessible by the respective identity provider. The work considers two types of models: basic and advanced versions. In the basic version, the cloud behaves as a service provider in the first phase. In the second, it authenticates the IoT device on behalf of the basic user. In the advanced version, the system provides the admin, cloud platform, and the manufacturer to access the IoT device for further maintenance activities.

The proposal offers a security methodology to secure the IoT network (Emerson, Choi, Hwang, Kim, & Kim, 2015). The authentication and authorization are the two main focus areas of that proposal. The authorization protocol uses OAuth protocol. The user permits the access facility to the security manager using the aid of the third party, and the security manager contains the database listing the record of the IoT devices. The user has to authenticate himself with the security manager before accessing the IoT device. The user ID evaluation aims in the completion of the validation procedure. The user access provisioned after successful validation.

The work suggests the effective implementation of the smart home system using IoT devices (Gaikwad, Gabhane, & Golait, 2015). The design can supervise activities and control the relevant appliances from anywhere. The system constitutes a GPRS module (Halonen, Romero, & Melero, 2004) and the radio frequency module (Song & Lee, 2003). The procedure begins with the user registering himself in the authentication page. The client module transmits its identity, password, and timestamp to the key distribution center for verification, and the transmitted data is encrypted using the secure hash algorithm SHA1 (Eastlake & Jones, 2001) or MD5 (Rivest, 1992). The authentication server collects the user profile. The application of advanced encryption standard (AES) (Sung, Kim, & Shin, 2018) to generate the ticket highlights the work. The ticket is transmitted to the user for monitoring and controlling activities.

The usage of one-time password authentication methodology provides better security (Shivraj, Rajan, Singh, & Balamuralidhar, 2015). The proposal uses elliptic curve cryptography and Lamport's OTP algorithm (Eldefrawy, Alghathvar, & Khan, 2011). The devices register themselves with the key generator, and the utilization of device identification for generation of public and private keys aims to secure the network. A time-variant generation happens when the application requests to transmit data. The input parameters include identity, time, counter and public parameters. Using these parameters the secret key calculation every session adds better safekeeping.

The RFID-based authentication protocol proposal (Fan, Gong, Liang, Li, & Yang, 2016) is a combination of different methodologies. The tag is used to store the credential hash function and is capable of generating a random number. The key and the random number usage are done in the generation of hash function. The procedure then aims to reduce computation and transmission costs. This methodology minimizes the occurrence of a denial-of-service attack, replay attack, eavesdropping, and spoofing in the network.

The time required to exchange messages aims to improve security (Sciancalepore, Piro, Boggia, & Bianchi, 2016). The proposed vital management protocol combines the functionality of implicit certificates with the elliptic curve Diffie–Hellman exchange (ECDH) (Kumar & Singh, 2016). The public coefficients of the two factors in consideration are exchanged. The nonces exchanged for every session enable better safety measures. The transferring done using two supplementary authentication messages follows the previous step, and an initial ECDH exchange following this procedure. The methodology authenticates and also aids in key generation.

A lightweight methodology (Esfahani, Mantas, Matischek, Saghezchi, Rodriguez, Bicaku, Maksuti, Tauber, Schmittner, Bastos, 2019) aims to secure Machine–Machine (M2M) communication (Kim, Lee, Kim, & Yun, 2014) in the industrial IoT environment (Hossain & Muhammad, 2016). The procedure suggested is based on hashing (Garg & Sharma, 2014) and XOR operations, and is is efficient against the replay attack (Mo & Sinopoli, 2009), a man-in-the-middle attack (Callegati, Cerroni, & Ramilli, 2009), impersonation attack (Ku & Chang, 2005) and modification attack (Yang, Hwang, & Lin, 2013). The sensor undergoes mutual authentication with the router (Chen, Tien-Ho, & Shih, 2010) and a registration procedure, before sending its identification to the server which then after verification generates and transmits hash code and a pre-shared key set using the secure channel. Using the received parameters the sensor mutually authenticates with the router. Using this methodology, the work characterizes low computation, storage and communication cost, and can achieve session key agreement and device identity confidentiality.

The device-free methodology is adopted to validate the users (Shi, Liu, Liu, & Chen, 2017). Some of the devices considered in the study include a laptop, smart refrigerator, microwave oven, and a printer. The system uses WiFi signals to depict personnel activities, and the behavior can be physiological or characteristic. Aggregation of the actions from the regular routines of the individual follow the procedure. The routines include stationary or dynamic behavior. An in-depth learning approach proposal detects an individual (Lecun, Bengio, & Hinton, 2015). The same procedure is employed to verfies the individual. The amplitude and relative phase of channel state information (CSI) is accumulated and evaluated, and this CSI aids in the generation of Orthogonal Frequency Division Multiplexing (OFDM) signals (Nee & Prasad, 2000). The signs carrying information of the personnel vary while sending data through multiple subcarriers, resulting in scattering, fading, and distortion, and alleviates using the relative phase. A subcarrier selection methodology is adopted that depicts the personnel characteristics. The proposed method captures six time-domain features and three frequency-domain features.

The work is based on physical unclonable functionality to validate the user (Aman, Chua, & Sikdar, 2017). The IoT device computation aims to communicate with the server and other IoT devices. The server issues a challenge Response Pair (CRP) for a device. Time based one-time password algorithm usage aims in generation of the CRP. The server stores the identification of the IoT device, and the device then communicates using its license with the server. On positive confirmation of the license, the server uses the random number and response of the IoT device, and using these parameters the encrypted message is generated. The transmission of the generated message, message authentication code and challenge of the device follows. The IoT device verifies the message using PUF. This procedure aids in mutually authenticating both the communicating devices, and the two communicating IoT devices then exchange identity information witheach other. They then generate CRP and an encrypted message.

A proximity-based methodology proposal (Zhang, Wang, Yang, & Zhang, 2017) focuses on securing the network. Utilizing the received signal strength, the devices are detected. The tools will have a significant amount of accepted signal strength variation when placed within the specified proximity. The observation consists of two differences: the difference observed due to the weakening of fast-changing channels, and antenna polarization used to bring significant changes between them. The device transmits a public key, and the procedure follows a series of private keys in encrypted form. The message embeds the MAC address. The Smartphone decrypts the message and validates the packet for its origin, and a random shared key is generated by the Smartphone and shared with the device.

Lightweight hash algorithms and XOR operations aims to secure the network (Dhillon & Kalra, 2017). The proposed work has four phases: user registration, login, authentication, and password change phase. Registration phase has two stages: the user is registered with the gateway, and sensors are then registered with the portals. During the login phase, the user authenticates using biometrics and a password. During the authentication phase, the user sends the verification message to the sensors. The procedure is performed securely by generating a one-time encrypted session key. The algorithm also includes a password change phase where the users can change the password.

Tewari & Gupta suggest a mutual authentication procedure (2017). The proposal consists of a tag with the reader and the server. The tag stores two sets of 96-bit tag identification and pseudonym. The tag and the server share critical information. The procedure provides the flexibility to use old fundamental benefits to complete the authentication procedure, and the evaluation of identity is performed by using XOR and left rotation operation. Bitwise addition modulo two is used to accomplish XOR operation.

A lightweight mutual authentication algorithm is suggested for smart city applications (Li, Liu, & Nepal, 2017). The proposed work consists of four stages: the security parameter consideration aims to generate a set of values in setup phase; these values are then used in the key generation stage to create public and private keys; the identities of the two communicating devices generate the public keys; then, using the same, they undergo mutual authentication.

A decentralized authentication system suggestion aims to secure IoT devices (Hammi, Hammi, Bellot, & Serhrouchni, 2018). The devices are zone-based. Only the members of the zone can communicate with each other. The public blockchain focuses to secure the systems (Bahga & Madisetti, 2016), and communication has to be validated by the blockchain. The master device is the vital distribution center. The work provisions devices willing to communicate with the elliptic curve private–public key-pair. The proposal provides tools with the tickets. The ticket is the aggregation of group identification, object identification, public address, public key, and signature. After the group creation the master device transmits the master identification. The blockchain validates the master identification and the group identification, and on positive affirmation the bubble creation follows. The method uses this as a ticket to proceed with its transactions.

The proposal (Kumari, Karuppiah, Das, Li, Wu, & Kumar, 2018) overcomes the shortcomings (Kalra & Sood, 2015b). The device undergoes the registration phase by transmitting its identity to the cloud server. The cloud server then verfies the identification and generates hash values, and these values are then used by the device during the login phase to authenticate it. The server and the device undergo mutual authentication and create the session key. The methodology minimizes the offline password guessing attack (Tsai, Lo, & Wu, 2013), insider attack (Schultz, 2002) and stolen-verifier attack (Chen & Ku, 2002).

The proposed work (Amin, Kumar, Biswas, Iqbal, & Chang, 2018) suggests a safe methodology to store data on private clouds. During the registration phase, the user generates the hash code using a random number and password. He forwards the same with the registration message containing the identity, hash code and password to the server. The server then calculates the hash code using status and a random number. The hash code obtained is concatenated with another random number to derive another hash code. The same is forwarded to the user securely. The hash code calculated by the user is then concatenated with identity to obtain yet another hash code. The hash code sent by the server utilization aims to concatenate with the hash code generated by the server. The smart card embeds this data, and the user uses the card by providing the identity and password. The server calculates the respective values using the hash code. The user admits on positive confirmation. The mutual authentication undergoes during the authentication phase. The key agreement phase follows the above procedure. The system also contains password and identity update phases.

Biometric technology is utilization (Hamidi, 2019) enforces security in the work. The technology creates unique identification and obstructs forging. The proposed work uses biometric technology to develop smart health monitoring in IoT. Biometric sensor utilization aids in collection of dat, and the data center collects the transmitted data. A private Weber is used to connect to smart and internet devices. The proposal utilizes a short-range communication system.

The asymmetric elliptic curve cryptography (Kumari, Abbasia, Kumar, Srinivas, & Alamd, 2020) utilization (Santoso & Vun, 2015) aims to secure the environment. The system uses WiFi gateway. This gateway provides a role to make initial configuration. This gateway is responsible for validating the devices for their identity. It also provides the flexibility to the user to set up, access and control the system using the equipment running the application program. After the completion of authentication methodology, both the communicating parties use the elliptic curve method. The network uses Diffie–Hellman methodology to generate a shared key, which is used by the the transmitted data as it undergoes encryption.

5.3 Proposed Work

5.3.1 Notations Used in the Study

Table 5.1 lists the notations used in the proposed action.

TABLE 5.1

List of Notations Used in the Proposed Work

Notations Used in the Study	Description
P_i	i^{th} patient
L_i	Location information
B_i	Biometric value
S	Server
$\eta(L_i)$	Hash code generated using location L_i
$\eta(B_i)$	Hash code generated using the biometric text B_i
B_p	Biometric reading of the personnel who is attending the patient
$\eta(B_p)$	Hash code of the biometric reading of the personnel

5.3.2 Assumptions Made in the Study

- The network considers the server as the most reliable device.
- It is the responsibility of the server to provide essential credentials to its client, authenticate them, and store the received data.
- The client devices are capable of generating the hash code using their location information.

5.3.3 Proposed Health Monitoring System

In the suggested system, the collection of biometric from the respective biometric devices enhances security. The updating of the data regarding the patient strengthens security. The methodology authenticates the server, followed by updating the corresponding data.

When the patient gets admitted to the hospital premises, his biometric is extracted. The work includes generation of hash codes using the same extracted output. In Equation (5.1) P_i is the patient, B_i is the biometric, and $\eta(B_i)$ is the hash code generated. The hash code generation implementation enhances security in the network (Lin, Jin, Cai, Yan, & Li, 2013). The updating of extracted data in the hospital server S follows.

$$P_i \rightarrow \eta\left(B_i\right)\|B_i : S \tag{5.1}$$

To initiate the update procedure, the staff member has to provide his biometric. The extracted data undergoes verification against the data stored in the server. In Equation (5.2) the biometric of the staff member is extracted to generate the hash code $\eta(B_i)$ and the same is transmitted to the server S.

$$B_p \rightarrow \eta\left(B_p\right) : S \tag{5.2}$$

The data collected undergoes verification against the server data. On positive confirmation, the server allows further update. The patient's data is updated by extracting the biometric, and the generation of a hash code follows. The hash code of the location obtained by the device follows. The same with biometric hash code undergoes concatenation and transmission to the server for an update. In Equation (5.3) the device is generating biometric hash code $\eta(B_i)$ by extracting the biometric readings from the patient P_i. The work

TABLE 5.2

Values Considered to Generate Hash Code

Input Parameters	Biometrics Reading	Location Information
Readings considered	*friction ridges of the skin consideration (length and angle) are the highlights of the text gathered.*	*(latitude and longitude values)*
No of values considered	10	1 for latitude + 1 for longitude
No of bits assigned	Four bits to represent each digit.	Four bits used to represent each digit.
Length of each reading	3 whole numbers + 1 decimal values (e.g., 163.3)	Latitude value considered – 0 to 90 Longitude value considered – 0 to 180

undergoes concatenation of the hash code of location information $\eta(L_i)$, device hash code $\eta(B_i)$ and data D_i. Table 5.2 summarizes the details of generating the hash code.

$$P_i \rightarrow \eta\left(B_i\right)\|\eta\left(L_i\right)\|D_i \tag{5.3}$$

5.4 Simulation

The biometrics of the personnel and the patient extraction generates the respective hash value and the same is stored on the server. Whenever the staff member examines the patient, he authenticates himself by giving the biometrics. The hash value for the same is generated and cross-verified with the data stored on the server. On positive confirmation, the personnel are allowed to update the evaluation summary of the patient. The same is updated using the hash value of the biometric extracted and location information of the patient. This value is concatenated with the data and submitted to the server. The methodology aims to authenticate and increase the reliability of the system.

The work employs NS2 to simulate. Table 5.3 lists the variables used in the work along with the description for them.

TABLE 5.3

Parameters Used in Simulation Work

Parameters	Description
Area under surveillance	200m × 200m
No of nodes considered (for the role of the patient)	10
No of nodes considered (as personnel)	2
Length of hash code of the biometric reading	32 bits
Length of data	256 bits
Length of hash code of the location information	16 bits
Simulation time	60 ms

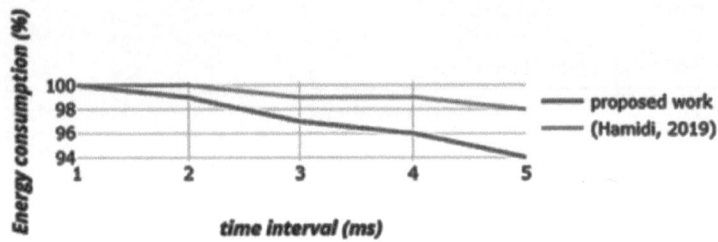

FIGURE 5.1
Comparison of energy consumption.

5.4.1 Energy Consumption

Energy is one of the essential resources. As these devices are to be reachable whenever necessary, security and reliability become the primary concern. The suggested proposal works to increase the safety of the data. The data provided has to be authenticated. In Hamidi (2019), the patient biometric is offered to update the readings. In some circumstances, it can breach the security of the system. The personnel entering the data may not be the authenticated one. Hence, to increase reliability, the proposed work considers using location information and personnel biometrics. Adding these parameters to the system increases the energy consumption on the IoT devices. The devices consume 2.016% more energy compared to Hamidi (2019). Figure 5.1 represents the energy consumption comparison.

5.4.2 Communication Overhead

The proposed work adds some extra parameters in the transmission of data to increase reliability to the system. The generation of location parameter is followed by hashing them. This parameter is concatenated with the hash value of the biometric reading of the patient and data to be transmitted, and the system also generates the hash value of the personnel to validate the staff. The proposed work has a communication overhead of 16.6% compared to (Hamidi, 2019). Figure 5.2 represents the same.

5.4.3 Reliability to Data

The data provided for updating has to be very accurate. The treatment to be followed later for the patient is dependent on the updated data. Any inaccuracy can make things worse for the health condition of the patient. The proposed work uses hash value during

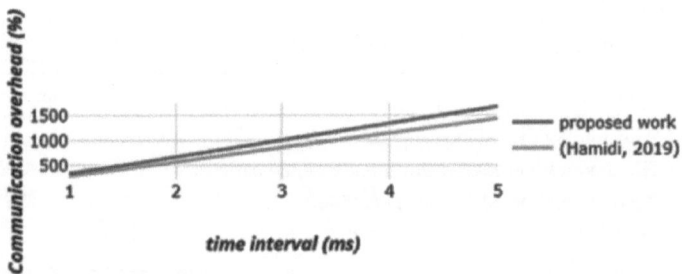

FIGURE 5.2
Comparison of communication overhead.

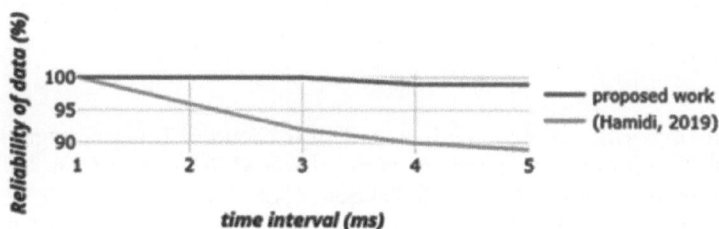

FIGURE 5.3
Comparison of reliability.

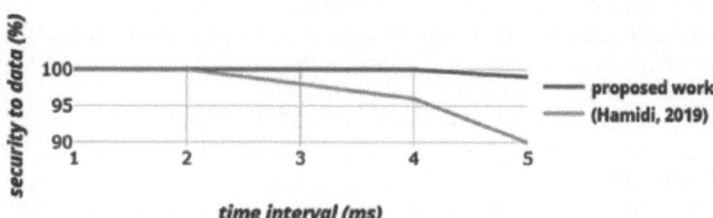

FIGURE 5.4
Comparison of security in the respective systems.

communication. This measure secures the data from many kinds of attacks. To provide more accuracy to data the hashed value of location information and personnel attending the patient are transmitted. Using hash codes decreases breaches and simultaneously adds authentication to the procedure. The proposed work increases reliability by 6.63% compared to Hamidi (2019). Figure 5.3 represents the same.

5.4.4 Security

Security is a necessity in the era of different kinds of attacks. The attacks can vary ranging from eavesdropping to manipulation of data. Appropriate measures consideration is meant to prioritize the patient's health. The proposed work uses hash codes to communicate with the server, and this increases security by 3.09% compared to Hamidi (2019). Figure 5.4 provides a graphical representation of the same.

5.5 Conclusion

IoT is providing convenience to humanity by enabling the tracking object of interest. These devices are programmed to transmit the sensed data to the respective using the internet. This facility aims at bringing the system under one roof, though they are miles apart. Security and reliability are vital issues to be considered. The proposal uses biometric text to enhance security in the network, and the gathering of biometric text of the personnel aims to increase security to the system. The hash readings of the biometric reading of patient and staff member along with their location are transmitted. The proposed work aims at enhancing security by 3.09% and reliability by 6.63%.

References

Aazam, M., Khan, I., Alsaffar, A. A., & Huh, E. N. (2014). Cloud of Things: Integrating Internet of Things with Cloud Computing and the Issues Involved. *International Bhurban Conference on Applied Sciences & Technology*. Islamabad, Pakistan: IEEE.

Alaba, F. A. (2017). Internet of Things security: A survey. *Journal of Network and Computer Applications*, 88, 10–28.

Alcaide, A., Palomar, E., Montero-Castillo, J., & Ribagorda, A. (2013). Anonymous authentication for privacy-preserving IoT target-driven applications. *Computers & Security*, 37, 111–123.

Aman, M. N., Chua, K. C., & Sikdar, B. (2017). Mutual authentication in IoT systems using physical unclonable functions. *IEEE Internet of Things Journal*, 4(5), 1327–1340.

Ambika, N., Mahmood Z. (eds) (2019). Energy-Perceptive Authentication in Virtual Private Networks Using GPS Data. In *Security, privacy and trust in the IoT environment* (pp. 25–38). Switzerland: Springer, Cham.

Amin, R., Kumar, N., Biswas, G. P., Iqbal, R., & Chang, V. (2018). A light weight authentication protocol for IoT-enabled devices in distributed Cloud Computing environment. *Future Generation Computer Systems*, 78, 1005–1019.

Andrea, I. C. (2015). Internet of Things: Security vulnerabilities and challenges. *IEEE Symposium on Computers and Communication (ISCC)*, 180–187. IEEE.

Arcelus, A., Jones, M. H., Goubran, R., & Knoefel, F. (2007). Integration of smart home technologies in a health monitoring system for the elderly. *21st International Conference on Advanced Information Networking and Applications Workshops (AINAW'07)*, Niagara Falls, Ont., Canada, 820–825. IEEE.

Atzori, L., Iera, A., & Morabito, G. (2010). The internet of things: A Survey. *Computer Networks*, 2787–2805.

Bahga, A., & Madisetti, V. K. (2016). Blockchain platform for industrial internet of things. *Journal of Software Engineering and Applications*, 9(10), 533.

Balasubramanian, A., Mahajan, R., & Venkataramani, A. (2010). Augmenting mobile 3G using WiFi. *Proceedings of the 8th International Conference on Mobile Systems, Applications, and Services*, 209–222. ACM.

Barreto, L., Celesti, A., Villari, M., Fazio, M., & Puliafito, A. (2015). An authentication model for IoT clouds. *IEEE/ACM International Conference on Advances in Social Networks Analysis and Mining*, 1032–1035. IEEE.

Callegati, F., Cerroni, W., & Ramilli, M. (2009). Man-in-the-Middle attack to HTTPS protocol. *IEEE Security and Privacy*, 7(1), 78–81.

Chen, C., & Ku, M. (2002). Stolen-verifier attack on two new strong-password authentication protocols. *IEICE Transactions on Communications*, 85(11), 2519–2521.

Chen, Tien-Ho, & Shih, W.-K. (2010). A robust mutual authentication protocol for wireless sensor networks. *ETRI Journal*, 32(5), 704–712.

Da Xu, L., He, W., & Li, S. (2014). Internet of things in industries: A survey. *IEEE Transactions on Industrial Informatics*, 2233–2243.

de Meer, H., Liedel, M., Pöhls, H. C., Posegga, J., & Samelin, K. (2012). *Indistinguishability of one-way accumulators*. University of Passau.

Dhillon, P. K., & Kalra, S. (2017). A lightweight biometrics based remote user authentication scheme for IoT services. *Journal of Information Security and Applications*, 34, 255–270.

Eastlake, D., & Jones, P. (2001). US secure hash algorithm 1(SHA1).

Eldefrawy, M., Alghathvar, K., & Khan, M. (2011). OTP-based two-factor authentication using mobile phones. *Eighth International Conference on Information Technology*, 327–331. IEEE.

Emerson, S., Choi, Y. K., Hwang, D. Y., Kim, K. S., & Kim, K. H. (2015). An OAuth based authentication mechanism for IoT networks. *International Conference on Information and Communication Technology Convergence*, 1072–1074. IEEE.

Esfahani, A., Mantas, G., Matischek, R., Saghezchi, F. B., Rodriguez, J., Bicaku, A., & Bastos, J. (2019). A lightweight authentication mechanism for M2M communications in industrial IoT environment. *IEEE Internet of Things Journal*, 6(1), 288–296.

Fan, K., Gong, Y., Liang, C., Li, H., & Yang, Y. (2016). Lightweight and ultralightweight RFID mutual authentication protocol with cache in the reader for IoT in 5G. *Security and Communication Networks*, 9(16), 3095–3104.

Gaikwad, P. P., Gabhane, J. P., & Golait, S. S. (2015). 3-level secure Kerberos authentication for smart home systems using IoT. *1st International Conference on Next Generation Computing Technologies (NGCT)* 262–268. IEEE.

Garg, P., & Sharma, V. (2014). An efficient and secure data storage in Mobile Cloud Computing through RSA and Hash function. *International conference on Issues and Challenges in Intelligent Computing Techniques*, Ghaziabad, 334–339. IEEE.

Halonen, T., Romero, J., & Melero, J. (2004). *GSM, GPRS and EDGE performance: evolution towards 3G/ UMTS*. John Wiley & Sons.

Hamidi, H. (2019). An approach to develop the smart health using Internet of Things and authentication based on biometric technology. *Future Generation Computer Systems*, 91, 434–449.

Hammi, M. T., Hammi, B., Bellot, P., & Serrhrouchni, A. (2018). Bubbles of Trust: A decentralized blockchain-based authentication system for IoT. *Computers & Security*, 78, 126–142.

Hankerson, D., & Menezes, A. (2011). *Elliptic curve cryptography*. USA: Springer.

Hossain, M., & Muhammad, G. (2016). cloud-assisted Industrial internet of things(iiot)-enabled framework for health monitoring. *Computer Networks*, 101, 192–202.

Hou, J. L., & Yeh, K. H. (2015). Novel authentication schemes for IoT based healthcare systems. *International Journal of Distributed Sensor Networks*, 11(11), 183659.

Kalra, S., & Sood, K. (2015a). Secure authentication scheme for IoT and cloud servers. *Pervasive and Mobile Computing*, 24, 210–223.

Kalra, S., & Sood, S. (2015b). Secure authentication scheme for IoT and cloud servers. *Pervasive Mobile Computing*, 24, 210–223.

Khanna, A., & Anand, R. (2016). IoT based smart parking system. *International Conference on Internet of Things and Applications (IOTA)*, Pune, India, 266–270. IEEE.

Kim, J., Lee, J., Kim, J., & Yun, J. (2014). M2M service platforms: survey, issues, and enabling technologies. *IEEE Communications Surveys & Tutorials*, 61–76.

Ku, W.-C., & Chang, S.-T. (2005). Impersonation attack on a dynamic ID-based remote user authentication scheme using smart cards. *IEICE Transactions on Communications*, 88(5), 2165–2167.

Kumar, S., & Singh, R. (2016). Secure authentication approach using Diffie-Hellman key exchange algorithm for WSN. *International Journal of Communication Networks and Distributed Systems*, 17(2), 189–201.

Kumari, A., Abbasia, M. Y., Kumar, V., Srinivas, J., & Alamd, M. (2020). ESEAP: ECC based secure and efficient mutual authentication protocol using Smart card. *Journal of Information Security and Applications*, 51, 102443.

Kumari, S., Karuppiah, M., Das, A. K., Li, X., Wu, F., & Kumar, N. (2018). A secure authentication scheme based on elliptic curve cryptography for IoT and cloud servers. *The Journal of Supercomputing*, 74(12), 6428–6453.

Lecun, Y., Bengio, Y., & Hinton, G. (2015). Deep learning. *Nature*, 521(7553), 436–444.

Lee, J. Y., Lin, W. C., & Huang, Y. H. (2014). A lightweight authentication protocol for internet of things. *International Symposium on Next-Generation Electronics*, Kwei-Shan, Taiwan, 1–2. IEEE.

Li, N., Liu, D., & Nepal, S. (2017). Lightweight mutual authentication for IoT and its applications. *IEEE Transactions on Sustainable Computing*, 2(4), 359–370.

Lian, S. (2009). Quasi-commutative watermarking and encryption for secure media content distribution. *Multimedia Tools and Applications*, 43(1), 91–107.

Lin, Y., Jin, R., Cai, D., Yan, S., & Li, X. (2013). Compressed hashing. *IEEE Conference on Computer Vision and Pattern Recognition*, Portland, OR, USA, 446–451. IEEE.

Mahalle, P. N., Anggorojati, B., Prasad, N. R., & Prasad, R. (2013). Identity authentication and capability based access control (iacac) for the internet of things. *Journal of Cyber Security and Mobility*, 1(4), 309–348.

Melis, A. P. (2016). Public transportation, IoT, trust and urban habits. *International Conference on Internet Science*, 318–325, 318–325.

Mo, Y., & Sinopoli, B. (2009). Secure control against replay attacks. *47th Annual Allerton Conference on Communication, Control, and Computing (Allerton)*, Monticello, IL, USA, 911–918.

Moosavi, S. R., Gia, T. N., Rahmani, A. M., Nigussie, E., Virtanen, S., Isoaho, J., & Tenhunen, H. (2015). SEA: a secure and efficient authentication and authorization architecture for IoT-based healthcare using smart gateways. *Procedia Computer Science*, 52, 452–459.

Nee, R., & Prasad, R. (2000). *OFDM for wireless multimedia communications*. Artech House Inc.

Nelson, J. (2013). Effective physical security. In J. Nelson, *Biometrics characteristics* (Vol. 4, pp. 255–256). Elsevier.

Nguyen, H., Mirza, F., Naeem, M. A., & Nguyen, M. (2017). A review on iot healthcare monitoring applications and a vision for transforming sensor data into real-time clinical feedback. *21st International Conference on Computer Supported Cooperative Work in Design (CSCWD)*, Wellington, New Zealand, 257–262. IEEE.

Petrov, V., Edelev, S., Komar, M., & Koucheryavy, Y. (2014). Towards the era of wireless keys: How the IoT can change authentication paradigm. *IEEE World Forum on Internet of Things*, Seoul, South Korea, 51–56. IEEE.

Porambage, P., Schmitt, C., & Kumar, P. (2014a). PAuthKey: A pervasive authentication protocol and key establishment scheme for wireless sensor networks in distributed IoT applications. *International Journal of Distributed Sensor Networks* 10(7)1–14.

Porambage, P., Schmitt, C., Kumar, P., Gurtov, A., & ylianttila, M. (2014b). Two-phase authentication protocol for wireless sensor networks in distributed IoT applications. *IEEE Wireless Communicatins and Networking Conference*, Istanbul, Turkey, 2728–2733. IEEE.

Rivest, R. (1992). The MD5 message-digest algorithm.

Santoso, F. K., & Vun, N. C. (2015). Securing IoT for smart home system. *International Symposium on Consumer Electronics (ISCE)*, Madrid, Spain, 1–2. IEEE.

Schultz, E. (2002). A framework for understanding and predicting insider attacks. *Computers & Security*, 21(6), 526–531.

Sciancalepore, S., Piro, G., Boggia, G., & Bianchi, G. (2016). Public key authentication and key agreement in IoT devices with minimal airtime consumption. *IEEE Embedded Systems Letters*, 9(1), 1–4.

Shi, C., Liu, J., Liu, H., & Chen, Y. (2017). Smart user authentication through actuation of daily activities leveraging WiFi-enabled IoT. *18th ACM International Symposium on Mobile Ad Hoc Networking and Computing*, 5. ACM.

Shivraj, V. L., Rajan, M. A., Singh, M., & Balamuralidhar, P. (2015). One time password authentication scheme based on elliptic curves for Internet of Things (IoT). *5th National Symposium on Information Technology: Towards New Smart Smart World*, 1–6. IEEE.

Song, J., & Lee, S. (2003). Patent No. 6,597,143.22. Washington DC, U.S.

Sung, B.-Y., Kim, K.-B., & Shin, K.-W. (2018). An AES-GCM authenticated encryption crypto-core for IoT security. *International Conference on Electronics, Information, and Communication (ICEIC)*, Honolulu, HI, USA, 1–3. IEEE.

Tewari, A., & Gupta, B. B. (2017). Cryptanalysis of a novel ultra-lightweight mutual authentication protocol for IoT devices using RFID tags. *The Journal of Supercomputing*, 73(3), 1085–1102.

Tsai, J., Lo, N., & Wu, T. (2013). A new password-based multi-server authentication scheme robust to password guessing attacks. *Wireless Personal Communications*, 71(3), 1977–1988.

Yang, C., Hwang, T., & Lin, T. (2013). Modification attack on QSDC with authentication and the improvement. *International Journal of Theoretical Physics*, 52(7), 2230–2234.

Yao, X., Han, X., Du, X., & Zhou, X. (2013). A lightweight multicast authentication mechanism for small scale IoT applications. *IEEE Sensors Journal*, 13(10), 3693–3701.

Zhang, J., Wang, Z., Yang, Z., & Zhang, Q. (2017). Proximity based IoT device authentication. *IEEE Conference on Computer Communications*, Atlanta, GA, USA, 1–9. IEEE.

6

Novel Method for Detecting DDoS Attacks to Make Robust IoT Systems

Sahil Koul, Rohit Wanchoo, and Farah S. Choudhary

CONTENTS

6.1 Introduction: Background and Driving Forces

A DoS attack is characterized by an attempt by a hacker to prevent authorized users from using the resources. An attacker may attempt to flood a network and thus reduce a legitimate user's capacity to access it, preventing usage of a service, or disrupt service to a specific system or a user (Felix Lau et al., 2000).

A Distributed Denial of Service (DDoS) differs in the aspect, that the attacker does not directly attack a victim. Instead he searches and hacks a various number of insecure computers, which are known as "zombies". These zombies then collectively form a botnet to perform a DoS attack on the victim.

There are many different types of DDoS Attacks that may be attempted on any victim (Garber, 2000; Douligeris and Mitrokotsa, 2004; Gibson, 2007; Gonsalves, 2007). But the packet-flooding attack is the most common type used (Galli, 2007). In this type of attack, an attacker sends a large number of Transmission Control Protocol (TCP), User Datagram Protocol (UDP) or Internet Control Message Protocol (ICMP) to the victim, (Peng et al., 2007), escaping from this attack is very difficult for the victim because of following two reasons. Firstly, the collection of zombies in the botnet is very large. This produces a huge rush of traffic which will eventually flood the victim. Second, the zombies also spoof their address under the attacker's influence. This makes it very difficult to trace back the attack traffic. Now if we try to understand that if the server is not being able to be accessed by the intended user how will the IoT device be robust. In this project, we will be discussing a method to prevent this type of DDoS attack traffic (Gordon and Loeb, 2006; Handley, 2007; Haris and Hunt, 1999).

Distinguishing a DDoS attack traffic (hereafter, called just as attack traffic), from a normal bursty legitimate traffic is a very difficult task. To achieve this, we are going to study the traffic history pattern incoming to the system and use this traffic history pattern to determine whether the traffic is an attack or legitimate bursty traffic. Also, we are going to use the AR time series model and chaos theory to achieve the same.

A time series is a data taken at discrete values of time. The data points are then indexed or listed or graphed in time order. There are various models by which a time series can represent different stochastic processes. We will be using the autoregressive model to process and predict the network traffic data (Vafeiadis et al., 2012).

Also, we will be using chaos theory to determine the state of the system. Chaos theory is a branch of mathematics that studies the dynamics of a system, which is very susceptible to initial conditions (and hence appear to be random systems). By studying the Lyapunov exponent of the system (that is the divergence of the predicted data from the actual data) we can tell whether the system is chaotic or not. If the system is chaotic, the Lyapunov exponent remains positive, which states that the data is chaotic/random. If Lyapunov exponent is negative or zero (both the cases will be discussed later briefly), the system is not in a chaotic state.

The method proposed can be summarized as follows: In the first phase, we analyze the incoming packet traffic to see whether packet traffic is giving a suspicion of a DDoS attack. If the first phase gives a suspicion about a DDoS attack, the second phase processes the traffic using the Network Anomaly Detection Algorithm (NADA), based upon the AR time series model and chaos theory.

The benefit of using this two-phase system as proposed in this paper is that first, the NADA algorithm cannot work all the time as it will waste computational resources as it requires to solve some mathematical equations, which may not be suitable for all devices (especially some low-end systems or an IoT device). Secondly, the network traffic might increase tremendously during a DDoS attack, that we might even not require a NADA algorithm to determine a DDoS attack. Just by observing the change in traffic we can declare a DDoS attack and take the needed countermeasures. On the other hand, sometimes DDoS attack might be difficult to judge by just traffic history analysis. Thus, our proposed method covers the best of both worlds.

This technique can, therefore, be used to make robust IoT systems (for convenience, we shall from here onward consider an IoT based ERP system).

6.2 The IoT and ERP Environment

Internet of Things (IoT) as we all know is a system of well-connected computing devices that work over the Internet in tandem with digital devices, sensors, or even people and animals that have a Unique Identification number known as the UID. The beauty of an IoT system is that it can transfer the message or so to call as data without having a human-to-human or a human-to-machine interaction. The data is produced in real-time and therefore is of great importance for many businesses and their allied areas in total. The IoT tools have not only enabled us with Smart homes, Voice-controlled devices, automated processing, security systems, etc. but also made a remarkable revolution in the business domain. In the business domain the contribution of IoT based systems can range from Data and Lead generation to smoother logistics and having better & prompt customer service support. As per data from the IEEE spectrum, the number of IoT devices has increased 31% year-over-year and reached a figure of 8.4 billion in the year 2017 and further estimates show that there will be at least 30 billion more connected devices by 2020. The market value of IoT globally is projected to reach around $7.1 trillion by 2020 (Internet of Things, 2020) which itself marks the importance of the IoT based devices in the current industrial context.

Now to ease up the process we from here on will be considering an IoT based Enterprise Resource Planning (ERP) tool to understand how the method that we propose can help in building a secure & robust system.

To get an idea about what an ERP tool is let's understand the basic need of all businesses. The basic need of all businesses is resource management, turning the resource into a profitable conversion along with maintaining good service quality & healthy customer relations (Why you should integrate IoT with ERP, 2020). At the most basic level this is what an ERP tool does for you automatically. An ERP tool integrates all the business processes in a streamlined manner and thereby reduces costs of resource utilization, analyses and interprets data to produce leads and customized insights to maintain higher profits without investing much on prospecting, and also caters to the need of replying to a customer in your absence through previous data sets present in the ERP tools memory maintaining a strong customer relation. Thus, we can say that an ERP works as a single stop for your business needs and acts as a standpoint where you can access all the business information on a real-time basis. Now since we know that an ERP system or tool mainly works on with data that is stored in its memory, to have an efficient ERP system we need to have a storage space known as a centralized database system and the access and connectivity to various data receiving points that may be an IoT operated device as shown in Figure 6.1

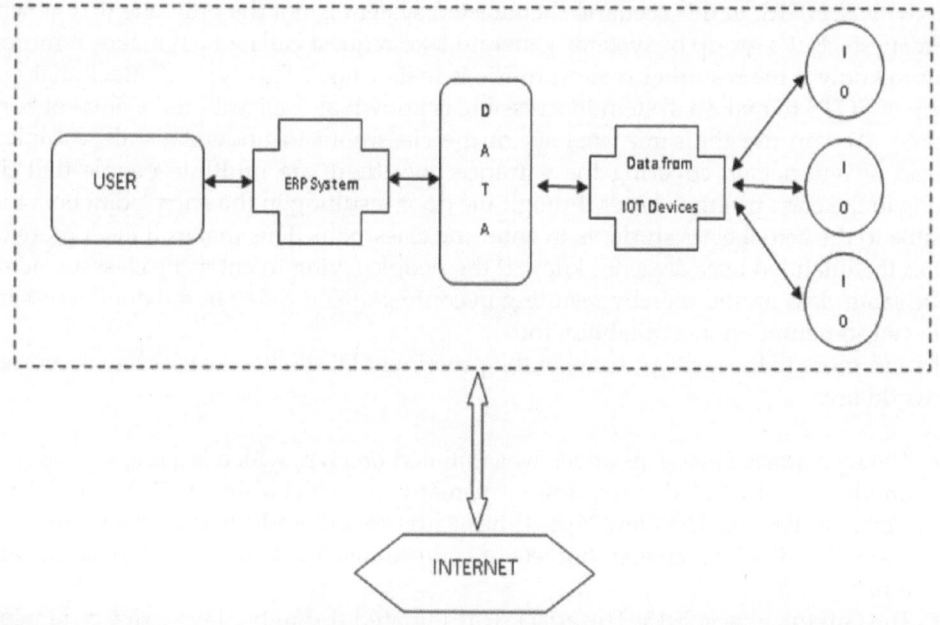

FIGURE 6.1
The ERP–IoT environment.

So, up until this point, we are clear that the whole thing relies on an efficient database and secured Internet to access that database. Now here's where the twist comes in the smooth process of having an IoT based device such as ERP's for your business process. Think of a situation in which you require some urgent data, and the ERP system cannot access the database due to heavy traffic on the tool. This will result in either a delayed or no-information fetch from the database; in either case the person accessing the information is at loss. But how can it be possible? The company that has chosen the ERP device must have had the consideration to know what & how much traffic will be on their ERP tool. Then how come the user doesn't get access to the system that has limited users acting on it? Here is where we talk about the term Denial of Service (DoS).

6.3 A Brief about Denial of Service (DoS) and Its Types

A Denial of Service (DoS) is a cyber-attack type wherein a malicious actor makes the system and its resources unavailable to the intended user by interrupting the normal functioning of the system. A DoS attack is generally triggered by flooding the network to be accessed by dummy requests, thereby hindering the normal traffic to use the network for the intended use and resulting in a DoS. A basic DoS attack is characterized by generating the requests from a single system. The infected system is then known as a BOT. The attack saturates the network with multiple requests that are of no use and thereby hinder the real traffic. This can be understood using the analogy of a person standing at the door of a classroom and not allowing the students to enter the class due to the inaccessibility of the entrance.

The DoS attack is further increased in its strength when there are multiple BOTS, or we can say BOT NETS. In this scenario the infected system is not the only one that generates the requests, but a group of systems generate fake request calls which makes it more difficult to know if the resulting request traffic is real or fake. This type of attack that uses a group of BOTS to make a system inaccessible is known as a Distributed Denial of Service (DDoS). We can use the same analogy of the classroom to understand this: think that instead of one person covering the entrance now there are multiple people that don't belong to the class trying to enter through the door resulting in the entry point being inaccessible to the actual class students to enter the classroom. This makes it even more difficult as the intended user does not know if the people trying to enter the class are actually of the same class or not, thereby resulting in confusion and delay in the door accessibility and even sometimes non-availability too.

To add more to the concept some of the prominent DDoS attacks that have occurred in the world are:

- The Dyn attack 2016: This attack was initiated on Dyn, which is a major DNS service provider, and created a disruption in many sites including AirBnB, Visa, Netflix, Amazon, PayPal, The New York Times, GitHub and Reddit. The attack used a malware (Mirai) which created botnets of compromised IoT devices to create the attack traffic.
- The GitHub attack 2015: The attack was initiated through a JavaScript code injected to the browsers that accessed the most popular Chinese search engine Baidu, which generated fake requests to the GitHub Pages.

- The Spamhaus attack of 2013: This attack was initiated on the Spamhaus which is an email spam combating agency. The attack created a traffic at a rate of 300Gbps and is considered to be one of the largest ever at the time DDoS attack.
- The Mafiaboy attack of 2007: An attack that was initiated by a 15-year-old hacker, by hacking the network of various universities to initiate the DDoS attack on various sites. The results were that it took down many major websites including Dell, CNN, eBay, and the most used search engine of that time, Yahoo.

(Famous DDoS Attacks, 2020)

These attacks have, however, caused the laws and regulations to become stricter, but still the attacks are not stopped; although they are reduced greatly through the ever-increasing research in the field of computing. In this chapter we too propose to design a novel method that can be helpful for creating secured and robust IoT systems. The method that we propose uses the network traffic history, NADA, concepts of chaos theory, AR series model, and related concepts to analyze the network traffic history and judge the suspicion of having a DDoS attack.

6.4 Related Work

Due to its distributed nature, a DDoS attack is very difficult to detect, as the origin of the traffic might be far away, or from various networks and different geographical locations. For this reason, DDoS is still a very powerful attack, capable of bringing down a network. Muhammad Aamir and Mustafa Ali Zaidi (2016) categorized a DDoS attack into various types of attacks. These are categorized as: network device level, OS level, application-level attacks, data flood attacks, and protocol level attacks, as shown in Figure 6.2.

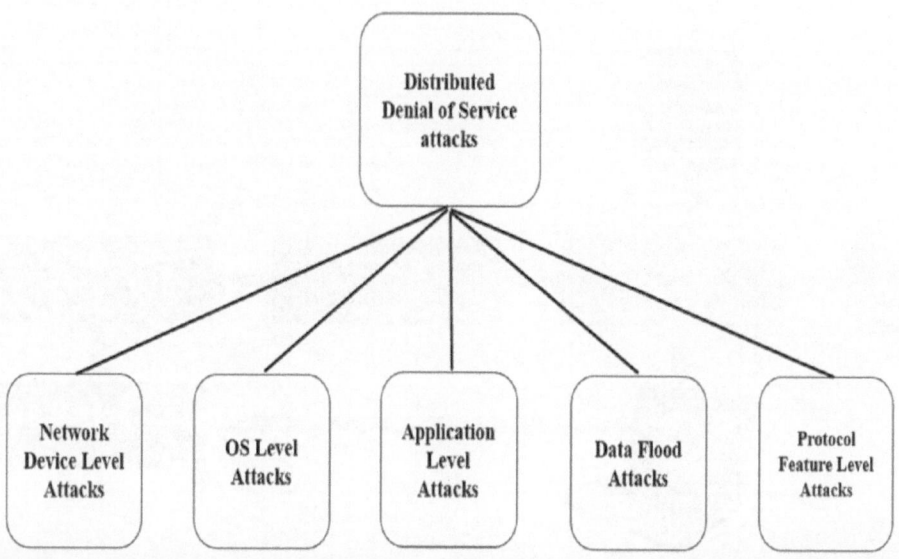

FIGURE 6.2
Types of DDoS attacks.

There are many measures that can be taken to prevent or stop DDoS attacks at various levels of the network. Stephen M. Specht and Ruby B. Lee (2004) mentioned many such detection and prevention techniques to tackle a DDoS attack. However, as these attacks are becoming more and more sophisticated, and due to the release of many attacking tools which allow any ordinary person to potentially perform a DDoS attack, these methods are fast becoming ineffective and outdated. To overcome these problems, Yonghong Chen et al. (2013) have developed a new method to detect DDoS attacks. This method uses the principles of time series modeling such as AR, ARMA, ARIMA, and FARIMA, etc. for analyzing and forecasting the network traffic. Then, by finding the prediction error and finding the Lyapunov exponent (using chaos theory), we can detect whether the incoming traffic is legitimate traffic or attack traffic.

6.5 Proposed Novel Method

The architecture shown in Figure 6.3 below of the proposed work is explained as follows:

1. The network traffic analyzer will constantly monitor the network traffic and will keep track of the incoming network traffic.
2. The analyzer will analyze the current network traffic and compare it with that of the network traffic history to see if there is any noticeably large increase in the network traffic.
3. Based on the comparative difference between incoming network traffic to that of the traffic history (stored in the traffic history database), the traffic analyzer will decide whether the traffic is a legitimate traffic or potentially DDoS attack traffic.

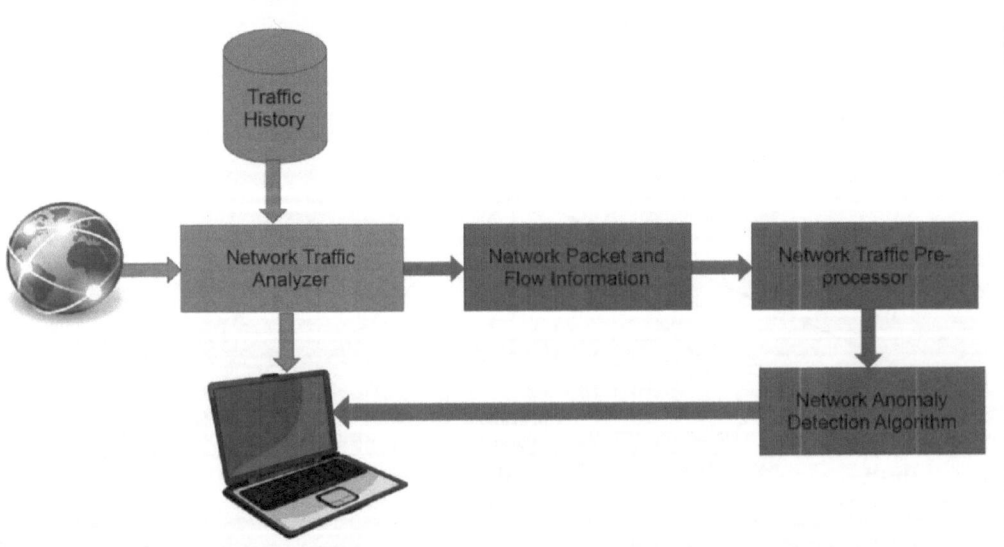

FIGURE 6.3
The architecture of the proposed method.

4. If the traffic analyzer determines that the traffic is legitimate, the process is stopped there, and the NADA algorithm is not used, saving the time and computing resources.

5. If the traffic analyzer is unable to signify that the traffic is legitimate, the traffic analyzer will pass control to the NADA (as shown in red in Figure 6.1) which will conduct further analysis on the nature of traffic patterns.

6. NADA first collects network packet and flow information and pre-processes it by cumulatively averaging the sequence, to suppress the network traffic.

7. Finally, we do a network traffic analysis to see if the traffic has any anomaly (i.e., whether the traffic is a burst of legitimate traffic, or a DDoS attack traffic) based on the AR time series model and Chaos theory, as shown in equations (ii) through (v).

8. If a DDoS attack is detected, the suspected packets are dropped until the traffic becomes normal again.

9. After this the packets will reach the server.

Let us suppose that in the starting our DDoS detection system has just been installed on the network server. The system will be fresh upon starting and will not have any history of the traffic on that network. We will then start to record the network traffic history at a regular interval of 15 minutes and store the traffic data along with the time of day and day of the week in the database. The traffic data will be stored as the number of packets that arrive at the server within these 15 minutes. The database table will be as shown in Table 6.1.

TABLE 6.1

Database Table

TIME	MON	TUE	WED	THU	FRI	SAT	SUN
00:00	345	520	586	578	800	790	850
00:15	350	500	600	620	750	800	1000
00:30	335	486	602	634	729	654	1050
00:45	351	492	590	640	744	635	1053
01:00	374	488	583	642	750	662	1044
01:15	342	477	569	626	743	521	1037
01:30	328	470	562	611	724	699	1021
01:45	326	440	500	601	700	708	800
02:00	300	380	458	582	683	717	749
02:15	314	356	402	570	664	688	812
02:30	286	340	490	509	670	644	899
02:45	294	300	468	515	687	631	967
03:00	273	280	400	518	644	622	932
03:15	286	290	389	494	605	634	901
03:30	249	280	387	422	555	642	876
03:45	264	274	350	387	514	604	867

The average field will eventually contain the average packets that a server receives on that particular day. After one week of installation, the system will have a complete history of network traffic for the previous week, and then we can start to analyze the network traffic.

To analyze the traffic, we will count the number of packets reaching the server for each 15-minute period. Then we will compare the traffic volume to the same day of the week and time of day as the current. This means that if the current network traffic has been monitored from 00:00 to 00:15 hours on Monday then we will compare it with the 00:15 column and row on Monday in the database. If the current traffic volume is either less, or the change is not more than a 10% increase in the traffic volume to the one we are comparing from the database, then we can safely say that the traffic coming to the server is legitimate traffic, and also if this value is larger than the previously recorded value, we will update the value in the database.

However, if the value exceeds the threshold limit of 10%[1], then such a traffic might be a DDoS attack. In this case we will analyze the change in the traffic values across the week passed. To do this we will find the change in traffic between the first two days and then store this value. Similarly, we will find the change in traffic values across all the consecutive days and store them as shown in Table 6.2 below.

Of these values, we will drop the negative values. Then we will find the average change in traffic across the week. This we will compare with the change in the traffic of the current day to that of the previous day.

For this system, one of three scenarios may arise:

1. If the incoming traffic is less than or comparable to the history of the traffic, the traffic can be safely declared as legitimate traffic. The server can safely take this load as we know it will have previously been able to take it.

2. If the incoming traffic is greater than the history traffic, we will analyze the incremental change in the traffic over the past week.

Let us understand this using an example: Consider the traffic values as mentioned in row 2 of Table 6.1. As we can see the traffic value stored in the database is 350 at 00:15 h. But over the course of the week, traffic at this time period is increasing, as is relevant from Table 6.1. Thus, it will be wrong to declare that an incoming traffic value of, say, 2,500 packets will be a DDoS attack. Thus, we will find the incremental change in the network traffic for this period, as depicted in Table 6.2.

TABLE 6.2

Data with Incremental Change in Traffic

Day	Traffic	Incremental Change in Traffic
Monday	276	
Tuesday	500	224
Wednesday	750	250
Thursday	1030	280
Friday	1500	470
Saturday	1790	290
Sunday	2060	270
Monday	2370	310

After this step, we will find the average of the incremental change, as shown below.

$$\text{Average difference of traffic} = (150 + 250 + 250 + 500 + 200 + 300)/6$$
$$= 297 \text{ which is comparable to } 310$$

As we can see from the above example, the average change in the traffic over a period of a week is comparable to the change in traffic volume on Monday. Hence, this will not be a DDoS attack.

3. If the traffic would not be increasing throughout the week, and will increase abruptly in a single day, it will most probably be a DDoS attack. In that case, we will detect whether this burst traffic will be a legitimate traffic or a DDoS attack traffic by detecting anomalies in the network traffic by NADA.

Now if we are not able to resolve whether the traffic is a DDoS attack or not (i.e., 3rd case occurs), we will use the method of traffic prediction. The basic rule of this method is that we predict the network traffic and then compare it with the original traffic value. This difference will be called as the prediction error. If we consider this prediction error as chaotic, then we can apply chaos theory to find the nature of traffic.

We use the AR time series model to predict the network traffic. However, in order to bring stability to the models, we sample the network traffic after collecting the network packets and flow information.

Let s_n be the different states of network traffic. Hence, we get a sequence as follows:

$$s_1, s_2, \ldots, si, \ldots, s_n \tag{6.1}$$

where s_i is the state of traffic to be predicted.

We can use the average of (6.1) over time period t_i to make the network traffic stable for accurate prediction, that is:

$$Z_i = (s_1 + s_2 + s_3 + \ldots + s_i) / t_i \tag{6.2}$$

Using autoregressive (AR) model, we can predict z_i:

$$V_j = \sum_{i=1}^{m} a_i Z_{j-i} \tag{6.3}$$

Hence, the sequence z_i can be generated from (6.2) and (6.3) as follows:

$$z_i = t_i Z_i - t_{i-1} Z_{i-1}$$

where z_i is the prediction of s_i

Hence, the prediction error can be found out as:

$$\Delta z_i = s_i - z_i$$

$$\rightarrow s_i = z_i + \Delta z_i$$

We now consider that the sequence $\{z_i\}$ represents normal traffic, whereas $\{\Delta z_i\}$ represents changed traffic due to additional, bursty legitimate or attack traffic.

Also, we will assume that $\{\Delta z_i\}$ behaves 'chaotically' when new traffic enters the system. By making this assumption, we analyze the mean exponential rate of $\{\Delta z_i\}$ which is the divergence between normal traffic $\{_i\}$ and the real traffic s_n. We can then use the Lyapunov exponent to observe change in traffic to see whether it is attack traffic:

$$L_i \approx \left\{ \ln\left[\Delta z_i/\Delta z_0 \right] \right\} / t_i$$

Now for the value of Lyapunov exponent, one of three cases arise:

Case 1: If $L_i > 0$, that is, the Lyapunov exponent is positive, the change in traffic is chaotic $\{ \Delta z_i \}$. This means that the change in traffic $\{ \Delta z_i \}$ is caused by new legitimate traffic entering the system.

Case 2: If $L_i = 0$, there is no divergence in the network traffic and the predicted network traffic. This means $\{ \Delta z_i \}$ is in a steady state that is, network traffic is constant.

Case 3: If $L_i < 0$, the $\{ \Delta z_i \}$ is steady and not random or chaotic (which it should be in case of a legitimate burst of traffic). This means that the change is caused by a DDoS attack traffic that may be introduced by an attacker affecting the system.

6.6 The Algorithm (Table 6.3)

TABLE 6.3

Algorithm Steps

Step 1	Get the volume of incoming traffic for a pre-specified period.
Step 2	Compare the current traffic volume with that of the traffic history.
Step 3	If the incoming traffic volume is determined by the traffic analyzer to be within safe limits we can safely say that there is no DDoS traffic present. We can stop here, and the below steps need not to be taken.
Step 4	Else, if the traffic analyzer is suspicious, we have to analyze the traffic by using NADA algorithm. For this, follow steps 5–9.
Step 5	Fetch the current packet flow information.
Step 6	Stabilize (pre-process) the network traffic with (ii).
Step 7	Predict the traffic with (iii) and (iv).
Step 8	Find the prediction error Δz_i and then detect any abnormal network traffic as explained.
Step 9	We can use this data to train neural networks to detect DDoS.

6.7 Implementation of Algorithm and Observations

We used Riverbed Modeler (Riverbed Modeler Academic Edition, version 17.5, Riverbed Technology, San Francisco, CA 94107, USA) to simulate three kinds of network traffic.

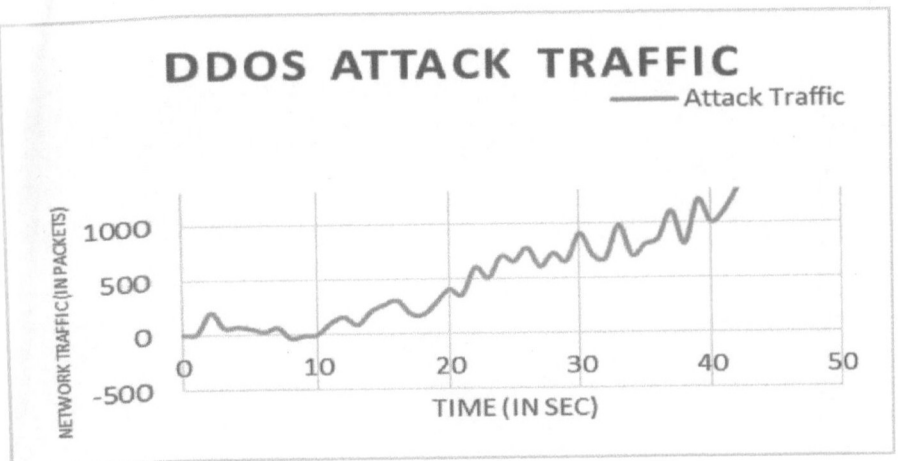

FIGURE 6.6
Graph showing a DDoS attack traffic.

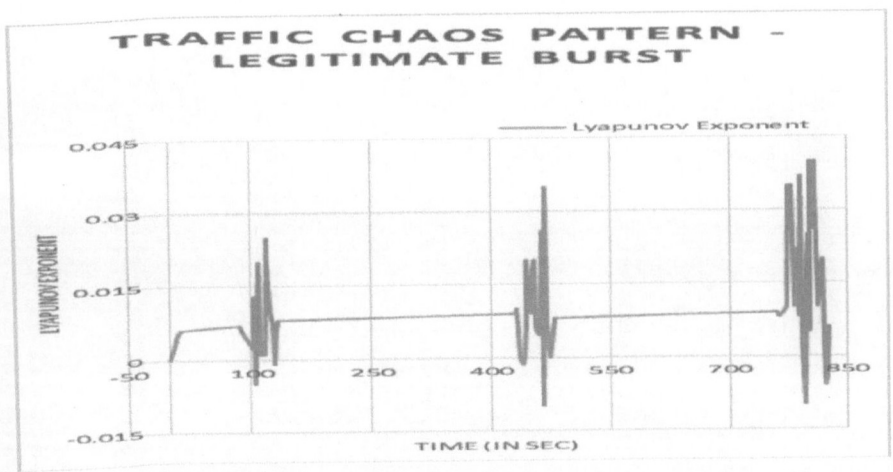

FIGURE 6.7
Graph showing traffic chaos pattern for bursty traffic.

value. So, we can say that the system is unpredictable or chaotic. Hence, we can safely say that the traffic is a normal legitimate traffic. Even if the Lyapunov exponent had become zero, then it would mean that there is no change in the divergence of the system, in other words there is no traffic entering the system.

The graph in Figure 6.8 shows the traffic chaos pattern for a DDoS attack traffic. The Lyapunov exponent Li is plotted against the time elapsed on the network. It can be observed that the Lyapunov constant is negative for most of the time. This means there is no divergence in the current and predicted traffic values of the system, which means that the traffic values in the system are predictable and not random or chaotic. Thus, the system has gone from a chaotic to a predictable state. This indicates that the incoming traffic is some value changing with some constant rate which is in case of DDoS attack traffic (Figure 6.6).

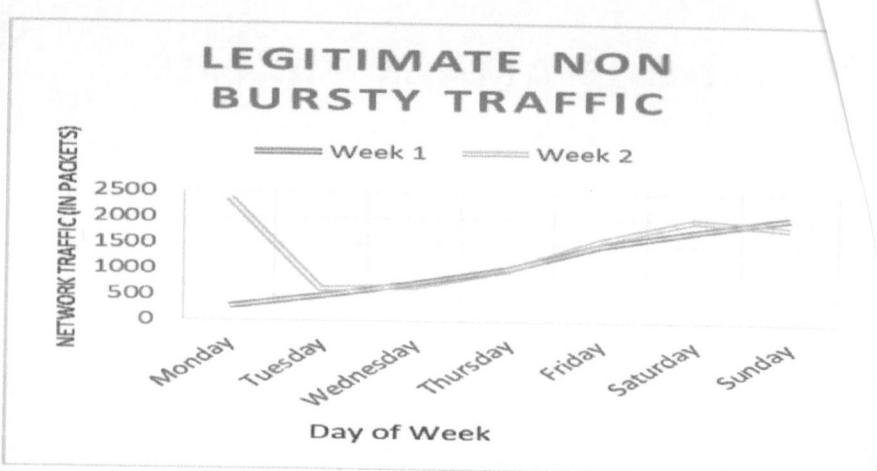

FIGURE 6.4
Graph showing non-bursty legitimate traffic.

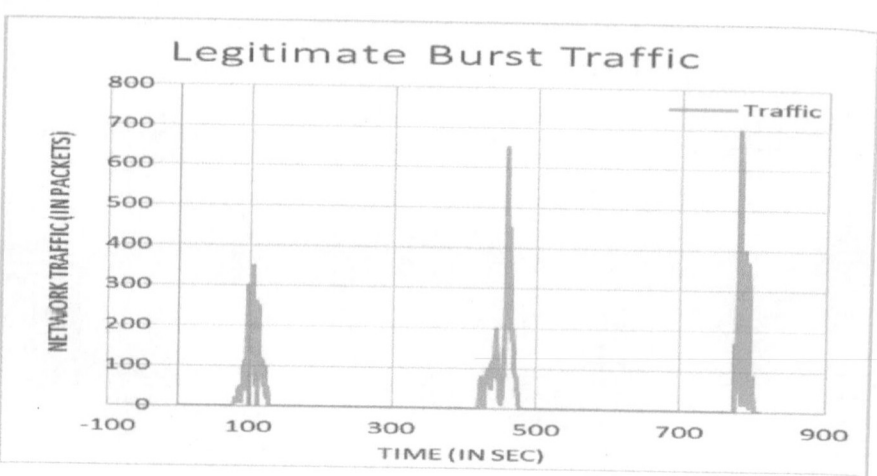

FIGURE 6.5.
Graph showing legitimate bursty traffic.

The first was the normal non-bursty traffic increasing over a week and then becoming somewhat bursty in nature. This kind of traffic did not require to use the NADA phase as it was simply resolvable by comparing it with the history as shown in Figure 6.4.

The second simulation was of traffic with a high-burst nature. This traffic varies rapidly over the period of time and cannot be compared using traffic history, as shown in Figure 6.5.

The third kind of traffic simulation was of a DDoS attack traffic as shown in Figure 6.6 which suggests DDoS attack traffic. This traffic goes on increasing over the time period, until it is either stopped or the server crashes.

The graph in Figure 6.7 shows the traffic chaos pattern for a normal legitimate burst traffic (Figure 6.5). The Lyapunov exponent Li is plotted against the time elapsed on the network. It can be observed that the Lyapunov exponent is positive for most of the time. This indicates that there is a divergence between the current traffic and predicted traffic

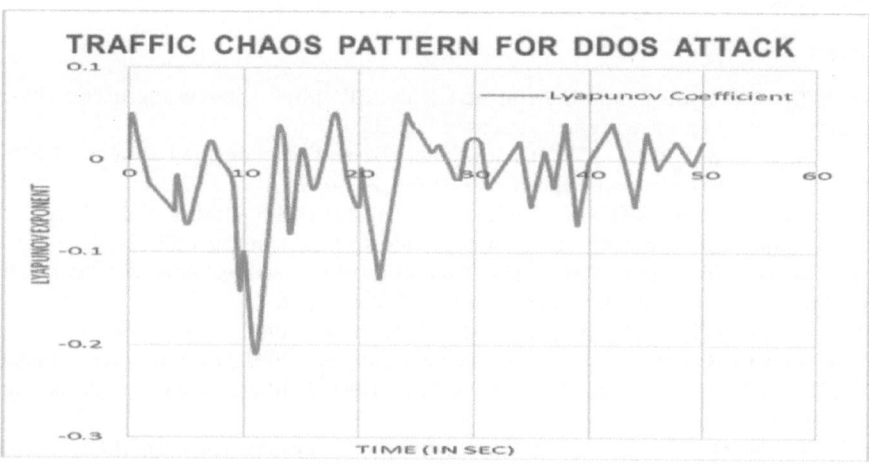

FIGURE 6.8
Graph showing traffic chaos pattern for a DDoS attack.

6.8 Conclusion and Future Scope

The method proposed in the text, despite being a somewhat novel way to detect the attack, is also vulnerable to some disadvantages which include:

- As the system takes some time to learn about the traffic patterns of a host, the detection of DDoS traffic can be unreliable initially.
- Full efficacy can be achieved by the system only when it is fine-tuned to a particular host's traffic pattern. Like the threshold traffic to engage NADA algorithm needs to be set and monitored for any particular system, till a balanced value is reached. It's not one blanket to all problems.

The method proposed in this paper is successful in detecting DDoS attacks. It is able to differentiate a DDoS attack based on the history of the incoming traffic and also by using prediction models and chaos theory. The method is much faster and more lightweight compared to the simple NADA algorithm, but can also detect bursty traffic nature and differentiate it from a DDoS attack.

We can use back-propagation trained neural networks to detect the DDoS attack by using the sample of abnormal traffic. That way we can highly increase the detection efficiency. We can also use a probability distribution function to tell the probability of a particular traffic volume on the day instead of just comparing the traffic volumes. This can further increase the detection efficiency of our model.

Note

1 The value of 10% is not fixed and can be adapted to suit different implementation procedures.

References

Allerin.com. 2020. "Why you should integrate IoT with ERP," https://www.allerin.com/blog/why-you-should-integrate-iot-with-erp. (accessed Jan, 2020).

Chonka, A., J. Singh, and W. Zhou. 2009. "Chaos Theory Based Detection against Network Mimicking DDoS Attacks," *IEEE Communication Letters*, 13(9), 717–719.

Cloudfare.com. 2020. "Famous DDoS Attacks | The Largest DDoS Attacks of All Time," https://www.cloudflare.com/learning/ddos/famous-ddos-attacks/, retrieved in 2020.

Douligeris, C. and A. Mitrokotsa. 2004. "DDoS Attacks and Defense Mechanisms: Classification and State of the Art," *Computer Journal of Networks*, 44(5), 643–666.

Felix, L., S.H. Rubin, M.H. Smith, and J. Trajkovic. 2000. "Distributed Denial of Service Attacks," in *IEEE International Conference on Systems, Man & Cybernetics 2000*, Nashville, Tennessee, USA.

Galli, P. 2007. "DoS Attack Brings down SUN Grid Demo," http://www.eweek.com/article2/0,1895,1941574,00.asp.

Garber, L. 2000. "Denial of Service Attacks Rip the Internet," *Computer Journal of IEEE*, 33(4), 12–17.

Gibson, S. 2007. The Strange Tale of the Denial of Service Attacks Against GRC.COM," http://grc.com/dos/grcdos.htm. (accessed March 12, 2019).

Gonsalves, C. 2007. Akamai DDoS Attack Whacks Web Traffic, http://www.eweek.com/article2/0,1895,1612739,00.asp (accessed March 16, 2019).

Gordon, A., P. Loeb, W. Lucysgyn, and R. Richardson. 2006. *CSI/FBI Computer Crime and Security Survey*, CSI Publications, Chennai, Tamil Nadu.

Peng T., Leckie C., and Ramamohanarao K., 2007. "Survey of Network Based Defense Mechanisms Countering the DoS and DDoS Problems," *Computer Journal of ACM Computing Surveys*, 39(1), 123–128.

Muhammad, A. and M. Ali Zaidi. 2016. DDoS Attack and Defense: Review of Some Traditional and Current Techniques, SZABIST, Karachi, Pakistan. https://docplayer.net/678758-Ddos-attack-and-defense-review-of-some-traditional-and-current-techniques.html (accessed May 2018).

Stephen M. Specht and Ruby B. Lee. September 2004. "Distributed Denial of Service: Taxonomies of Attacks, Tools and Countermeasures," in *17th International Conference on Parallel and Distributed Computing Systems, 2004 International Workshop on Security in Parallel and Distributed Systems*, San Francisco, California, 543–550.

Vafeiadis, T., A. Papanikolaou, C. Ilioudis, and S. Charchalakis. 2012. "Realtime Network Data Analysis Using Time Series Models," *Simulation Modelling Practice and Theory*, 173–180.

Yonghong Chen, Xinlei Ma, and Xinya Wu. May 2013. "DDoS Detection Algorithm Based on Preprocessing Network Traffic Predicted Method and Chaos Theory," *IEEE Communications Letters*, 17(5), (1052–1054).

Wikipedia. 2020. "Internet of Things," https://en.wikipedia.org/wiki/Internet_of_things (accessed Jan, 2020).

7

Machine Learning for Enhancement of Security in Internet of Things Based Applications

Sparsh Sharma, Faisal Rasheed Lone, and Mohd Rafi Lone

CONTENTS

7.1 Introduction

The drastic advancement in communication technologies such as the IoT has remarkably revolutionized living habits and is a further step toward modernization, leading to improved quality of life. IoT technology enables the devices present in our surroundings to communicate and exchange information with each other for working as per our requirements and needs. IoT is one of the quickly emerging technology, and it is estimated that in the coming years, every device will be equipped with some kind of intelligence. IoT has a plethora of applications, including smart healthcare, smart homes, medicine, ITS, smart education, etc.

IoT involves communication among the heterogeneous type of devices leading to some serious challenges which need to be addressed. Among such challenges, security is one important aspect that needs to be handled for a smooth, error-free and secure smart environment. Due to the involvement of various heterogeneous devices, there is a chance of the

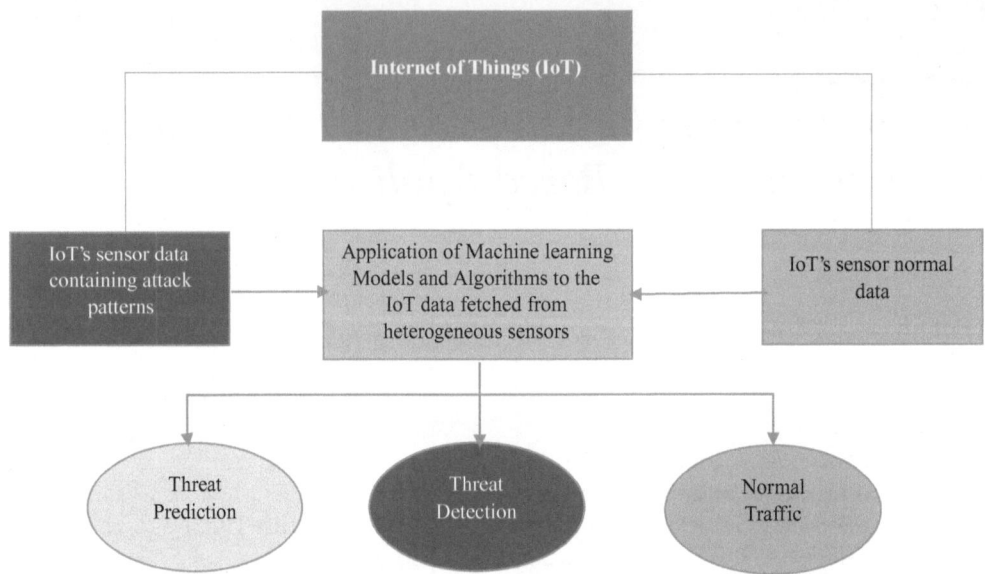

FIGURE 7.1
Generic model of applicability of machine learning to IoT network for threat detection.

existence of various security vulnerabilities and loophole, which can lead to drastic results like hindrance of privacy, confidentiality, and physical damages.

Machine learning has revolutionized the whole world with the plethora of applications supported by it and its contribution in enhancing the overall performance and intelligent decision-making. Machine learning has already proved to be very efficient in enhancing the security as well as detection of threats in various other networks like Vehicular ad hoc Networks (VANETs), Mobile ad hoc Network (MANETs), cloud computing, etc. In this study, the possibility of exploring the usage of machine learning in enhancing the overall security of the IoT network is examined. A generic model depicting the usage of machine learning algorithms for the separation of malicious traffic from the normal IoT traffic is shown in Figure 7.1.

7.1.1 Main Contributions of This Proposal

Key contributions of this study for security enhancement in IoT using the machine learning models, also shown in Figure 7.2, are as follows:

1. Comprehensive study and discussion on the possible security threats and challenges in IoT are thoroughly presented.
2. Drawbacks and limitations of existing security solutions that have led to move toward machine learning and deep learning-based security solutions have been studied and are discussed.
3. An in-depth analysis of machine learning and recent advancements in deep learning methods for IoT security has been done. Various machine learning and deep learning-based algorithms and solutions capable of securing the IoT network are studied and are presented with their advantages, disadvantages, and applications in IoT security. Along with that, comparisons of machine learning and deep learning from IoT perspective, with the help of tables and figures are presented.

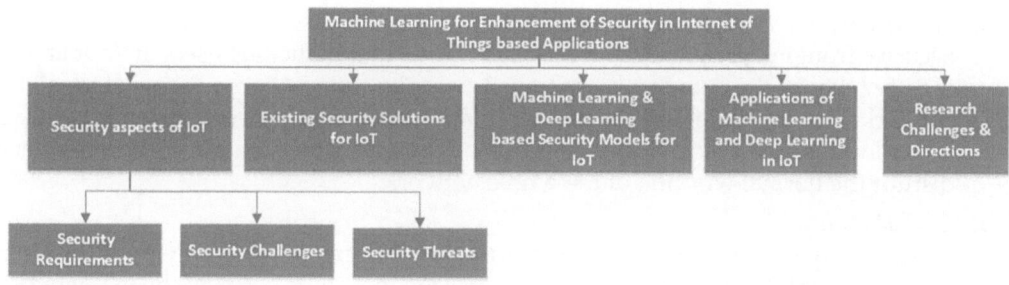

FIGURE 7.2
Blueprint of chapter.

4. Applicability of machine learning and deep learning in different applications of IoT has been explored and is presented.
5. Research challenges and future directions in the field of IoT security using machine learning and deep learning is enumerated and provided to help the budding researchers working on this particular domain.

7.2 Existing Security Solutions and Challenges Associated with Them

As mentioned above, security is a serious issue in IoT based networks which needs to be addressed. There exist various security solutions and techniques which are already being used in a variety of other networks. These security techniques include the use of encryption, cryptographic algorithms, Intrusion Detection Systems (IDS) (Sharma and Kaul, 2018a, 2018b), and many other application and network security-based tools and techniques. However, due to some inherent characteristics and features of IoT network like constrained resources, heterogeneous devices, dynamic changes of the state of the network, the enormous scale in which any number of devices can connect and communicate, etc., has made the application of those traditional security techniques inefficient for IoT.

So, in order to handle the security aspect of IoT, machine learning and deep learning solutions are gaining popularity. The machine learning and deep learning models, because of their capabilities in dealing with a variety of real-life problems, can be used alongside IoT for enhancing overall security. These models have added advantages over other existing techniques, like their capability in dealing with zero-day attacks.

7.3 Security Aspects of IoT

7.3.1 Security Requirements

Every network has some standard security goals that need to be satisfied to prevent it from being compromised. In this section, security goals like confidentiality, integrity, availability, authentication, etc., crucial for IoT based networks are presented.

a) *Integrity*

Security from integrity violation attacks ensures that malicious users are not able to modify/change the data during its transit. Security from integrity is crucial for all the networks, especially IoT where the information exchange between the smart devices is continuous, and any modifications during their transmission can compromise and disrupt the normal working of the whole network.

b) *Confidentiality*

This security goal ensures that the information that can be seen only by those users for which it is intended and not by unauthorized malicious users. In the IoT network, confidential information transmission is frequent and usually contains some sensitive information that should not be disclosed.

c) *Availability*

This security goal ensures that the information and resources are there for the intended users round the clock. The resources which can be anything depending upon the type of the application and network in use are accepted to be available to its intended users whenever they need it. Some of the security attacks which target this security goal of availability are Denial of Service (DoS) Attack, Distributed DoS Attack, Packet Dropping, Selective Forwarding (Sharma and Kaushik, 2019), etc.

d) *Authentication*

This goal ensures confidentiality, such that the sensitive information is accessed by only those users for which it is intended by making use of authentication and verification mechanisms. The authentication goal of security is used to provide and ensure proper authorization in the IoT network.

7.3.2 Security Challenges in IoT

IoT involves the integration of a variety of heterogeneous networks, and this integration may lead to various challenges which need to be identified and addressed before the actual deployment of the network to prevent any type of future damage and disaster. In this section, various security challenges that are prevalent in IoT are presented.

IoT involves the integration of various devices in surrounding to the Internet so that they can be accessed from a remote location only. The devices in the IoT environment usually have different characteristics and features. Also, the devices in IoT usually work unattended and in isolation. These devices have a high chance of being physically attacked and compromised.

7.3.3 Security Threats

This section focuses on the plethora of security threats possible in IoT based networks. These threats include attacks like DoS, Blackhole, Information Disclosure, etc., and can be categorized based on different parameters like active vs passive, insiders vs outsiders, availability vs confidentiality vs integrity, etc.

Although a wide range of work has been done in enhancing the security of the IoT network as well as to minimize the impact of these security threats in literature, the IoT network is still susceptible to security threats and attacks, and this needs to be addressed. So, in order to deal with these security threats, it is important to identify all the possible and prevalent security threats and attacks in IoT along with their mode and technique of operations.

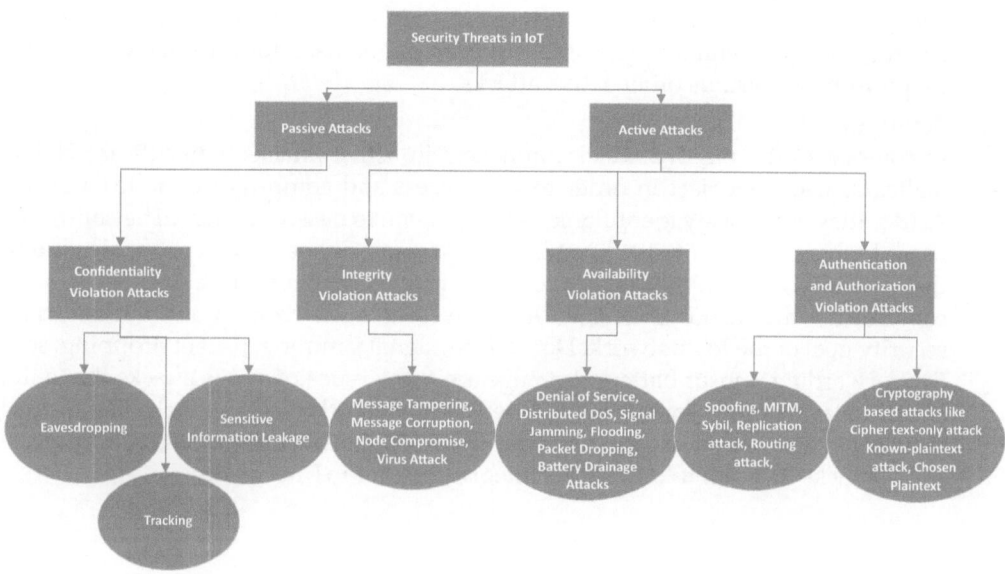

FIGURE 7.3
Categorization of security threats in IoT.

Security threats in IoT based on the method of compromising the IoT network can be categorized as physical threats and cyber threats (Al-Garadi et al., 2018). Physical Threats work by physically compromising IoT nodes, while cyber threats are usually performed remotely over the Internet. Cyber threats can be further categorized as passive attacks and active attacks. Various possible categorization of security threats in IoT network is shown in Figure 7.3.

7.3.3.1 Physical Attacks

Physical attacks usually work by physically compromising and destructing the IoT device, which can be an IoT Server or an IoT device (sensors or camera). This category of attack doesn't require any type of technical proficiency to conduct. In this kind of attack, all the reachable physical IoT devices are prone. It is quite visible, easy to perform, and is widely scalable because the majority of the physical devices and objects of IoT are present everywhere and are physically accessible. Physical damage to the IoT devices is also possible due to unintended natural calamities like floods, earthquakes, tsunami, etc.

7.3.3.2 Cyber-Attacks

a) *Passive Attacks*
   In passive attacks, the attacker passively gathers the sensitive network information by eavesdropping the network. This type of attack is difficult to detect and mitigate, because the attacker doesn't actively inject any malicious packet or information into the network to acqu24ire the network's information. The passive attack works by eavesdropping the communication channel or network information. Making use of cryptographic techniques and encryption of the traffic is one way to deal with passive attacks. A passive attack is usually an attack on the confidentiality goal of the

network and can be used to extract sensitive information like a patient's medical records, login credentials, etc. Passive attacks can be used for information gathering for performing various other active attacks.

b) *Active Attacks*

In contrast to the passive attacks, attackers in active attacks explicitly inject some malicious code or object in order to gain access and compromise the IoT network. Active attacks are easily identifiable in comparison to passive attacks. They are widely used. In this category of attack, malicious users not only eavesdrop, but also modify the network to disseminate false information for their own benefit. Active attacks can compromise the integrity, availability, as well as authorization and authentication security goal of the IoT network. DoS, DDoS, signal jamming, packet dropping, selective forwarding, sensor battery drainage are a few types of active attacks that aim to affect the availability goal of the IoT network. Similarly, active attacks like message tampering, injection of malicious codes, viruses, trojans, message destruction, etc. are a few attacks that aim to impact the integrity goal of IoT networks.

7.4 Existing Machine Learning-Based Security Solutions for IoT along with Their Limitations and Challenges

This section will do an exhaustive survey on the existing security solutions proposed for each attack. It will also focus on the advantages, disadvantages as well as simulators used in each proposed solution.

For dealing with the above-mentioned active as well as passive, both physical as well as cyber-attacks are possible in an IoT network, and various security solutions and frameworks have been proposed by researchers across the globe. However, each of them has its own advantages as well as limitations. Machine learning, which has proved to provide excellent performance and efficiency in almost all possible domains, can also be deployed in the IoT network for improving and enhancing its security. In this section, various machine learning models that can be used, deployed in the IoT network for securing the IoT network has been presented along with their advantages, limitations, and challenges, if any, in their deployment.

7.4.1 Convolutional Neural Network (CNN) for Security Enhancement of IoT

CNN is a deep learning algorithm that has been providing support and results in various research areas. The main advantage of CNN is its ability in learning from vast feature hierarchy and from a voluminous amount of unlabeled IoT data which is quite often in the IoT network. Extraction and classification are the two crucial phases of a CNN network. The extraction phase involves extracting relevant features for learning from data in raw form. The extraction phase is done by two basic layers of CNN, i.e. convolution layer that is responsible for the collection of data from the neurons, while another layer is the max pooling layer that follows the convolution layer and classifies the collected data to the sub-samples. The learning in CNN is done through backpropagation which involves repeatedly iterating to Convolutional and Max Pooling layers. Another vital phase of CNN is classification, in which the data is fed to the classifier and includes only one layer (i.e., fully connected layer that try to find the final solution by mapping of neurons to the final output).

For improving the performance of the IoT Risk Assessment, a CNN based model for IoT has been developed by Abbass et al. (2018), who have tried to address the various SRA challenges in the article by making use of the CNN based deep learning algorithm. The main objective of this article is to enumerate and predict security risks in IoT networks. In the proposed CNN based model, two sensors were used in which one is made to act as a client while others as a server. The raw data collected contains patterns of 14 attacks. Limitations: Data collections is time-consuming as well as computationally expensive. Future scope: Adding more attack samples to the dataset for improving the model's performance and accuracy.

Another work using Deep Neural Network (DNN) for the identification and classification of attacks in IoT network is presented by authors in (Tama et al., 2017). For evaluating the performance of the proposed DNN based security solution, various validation techniques like cross-validation, repeated cross-validation, and sub-sampling on different datasets have been performed. Based on the experimental results that were obtained after the implementation of DNN based model, it was found that DNN performed well when applied on the CIDDS-001 dataset no matter which validation technique was used and this could be due to the imbalanced issues in the dataset. *Dataset used*: UNSW-NB15 (Moustafa and Slay, 2015), CIDDS-001 (Ring et al., 2017), and GPRS (Vilela et al., 2014). *Future scope*: To conduct and investigate larger values of the experimental repetition.

7.4.2 K-Nearest Neighbor (KNN) for Security Enhancement of IoT

For dealing with the security attacks and to safeguard the IoT network, authors in (Pajouh et al., 2016) have proposed an anomaly-based IDS that uses dual-layer dimensional reduction methods as well as dual-tier classification module. Linear Discriminate Analysis (LDA) and component analysis are the methods that are used by authors for dimensionality reduction for extracting only the useful features from the high dimensional datasets in order to get rid of redundant features. For the classification of malicious users having suspicious behaviors, Naïve Bayesian and KNN methods are used. *Attacks addressed:* User-to-Root (U2R), Remote-to-Local (R2L). Dataset: NSL-KDD (Lakhina et al., 2010). *Accuracy:* Detection Rate = 84.82%, *False Alarm Rate:* 5.56%. *Future scope:* Exploring the possibility of using fuzzy clustering against attacks like U2R, R2L, and other attacks.

Another KNN model is brought in use by authors (Guo et al., 2019) for handling the uncertain data stored in semi-trusted cloud servers. For encryption of uncertain IoT data, homomorphic encryption involving two servers has been used. A safe and efficient KNN search model is used by making use of the expected rank over the encrypted data. The main focus of the authors was to maintain the security of the data while improving the efficiency of the query. Dataset: U.S. Census Bureau (BOC, 2014).

7.4.3 Support Vector Machines (SVMs) for Security Enhancement of IoT

SVM is one of the powerful machine learning classifiers being brought in use in various applications of different domains because of its high performance. SVM can be used to solve and classify both linear as well as nonlinear data easily. SVM removes unnecessary and redundant features on its own while carrying out the modeling and classification. SVM works by finding an optimal hyperplane that separates the different classes and is the criteria for the learning process in order to classify the data accurately. In SVM, out of the various available hyperplanes, SVM uses the hyperplane that maximizes the margin (i.e., the distance between the data, and thus, in turn, maximizes the generalization capability).

Ham et al. (2014) have made use of Linear SVM for the detection of malware in android-based phones, as android phones are usually brought in for the use of accessing and monitoring the IoT devices. The SVM detects the malware of the smartphone by considering the data collected by monitoring the resources on the android device. For checking and comparison of the performance of the proposed SVM-based android malware detection framework, other classifiers like decision tree, Naïve Bayesian, Random Forest were used. In contrast to the use of 88 features in previous work by Shabtai et al. (2012), they had used only 32 features, which can be extracted even without the root permission. Features that were considered by authors in that work were categorized in the following types: network resources, telephone related, SMS, CPU, battery, process, and memory. *Advantage:* The features that were brought in use for the malware detection had been reduced to just 32, in comparison to the 88 features used by Shabtai et al. (2012). All the features can be extracted without the root permission. *Accuracy:* True positive: 99.99%, false positive rate (FPR): 0.004, Accuracy: 99.5%. *Future scope:* To consider the detection of hardly detectable android malware using resource information with higher accuracy. Deployment of lightweight SVM-based detection module on android for real-time detection of malware.

Another security framework using a multi-classification algorithm based on double SVM decision tree has been proposed by Li and Applications (2020). In this research article, authors have proposed a location privacy security mechanism that is based on anonymous tree and box structure, which provides location privacy protection support for services dedicated to intelligent terminals. The authors have compared their proposed system with MBSVM and 1-V-1 SVM for evaluating and comparing the performance. Based on the performance results, it has been found that this technique shows a reduction in time bandwidth overhead as well as improving the node location consistency. *Dataset:* KDDCup99 (Tavallaee et al., 2009). *Attacks:* DoS, U2R, R2L, Probe. *Future scope:* Use of lightweight cognitive technology to maintain the credibility of the user's identity.

For ensuring secure and reliable data sharing in IoT, a blockchain-based security mechanism is used by Shen et al. (2019). A combination of Paillier cryptosystem and the blockchain-based mechanism is being used to handle the attacks against data privacy, ownership and integrity while using the SVM classifier for training using the IoT data. Firstly, the data in the IoT System is encrypted using the Paillier cryptosystem, which is then stored in a blockchain-based distributed ledger. Using the encrypted data obtained by communicating with the corresponding data provider, the SVM classifier is trained. It pertains to mention that the plaintext is never provided for the training purpose. For the training of classifiers on the encrypted data, secure polynomial multiplication and secure comparison are the two security protocols that were used. *Datasets used:* Breast Cancer Wisconsin Data Set (BCWD) and Heart Disease Dataset (HDD) (Detrano et al., 1989). *Accuracy:* Precision: 90.35 % for BCWD and 93.89% for HDD dataset. *Recall:* 96.19% for BCWD and 89.78% for the HDD dataset. *Future scope:* To develop a generalized security framework that allows the construction of privacy-preserving machine learning training algorithms for encrypted datasets.

In order to prevent the existing issues of handling heterogeneous data types in various security protocols designed for IoT, Sankaranarayanan and Mookherji (2019) made use of the SVM classifier of machine learning for analyzing the traffic pattern in IoT network and to detect the traffic anomalies. The SVM-based algorithm has been applied to the traffic data set of three cities of the United Kingdom from the period 2011–2016. For implementation and demonstration of the capability of the SVM algorithm in the classification of traffic as good and bad, Raspberry PI3 as an edge-level device has been used. *Accuracy:* 87%, Precision-Recall: 86%. *Attack:* Man-in-the-Middle. *Future scope:* Developing a system to deal with unknown attacks and using unsupervised learning-based techniques. Accuracy can be improved by using more amount of traffic data. Also, deep learning-based methods can be deployed to deal with known and unknown security attacks.

7.5 Applications of Machine Learning and Deep Learning in IoT Based Network

Apart from security, various areas of IoT where machine learning and deep learning-based solutions can be used are discussed in this section. Machine learning- and deep learning-based models have provided aid to various applications of IoT improving and enhancing its overall performance and efficiency. Some of the areas of IoT where machine learning and deep learning models can be used are:

1. *Traffic Management and Routing*: Combination of machine learning and IoT can be used for the efficient management of road traffic and routing of vehicular traffic. The sensory road and traffic data can be fed to the various machine learning algorithms for various efficient and smart decision-making for preventing road congestion, accidents and for quick traffic dissemination.

2. *Smart Homes:* With rapid advancements in IoT, Smart home systems are gaining popularity in our daily life. Smart homes using IoT adds convenience, functionality, and efficiency for people. However, despite providing lots of benefits, it also exposes humans to various security as well as privacy risks. So, machine learning-based techniques can be used not only to eradicate the security loopholes of the smart homes, but also to make smart decisions based upon the person`s daily usage.

3. *Healthcare:* IoT and machine learning-based algorithms go hand-in-hand with enhancing each other's capabilities and performance. IoT works on the voluminous data fetched from various sensors. At the same time, machine learning algorithms can be brought in use for extracting and making intelligent decisions based upon the requirement and situation. Various combined applications of machine learning and IoT in the healthcare sector include:

   - Inventory tracking of patients, doctors, and medical staff
   - Drugs and medicine management
   - Waiting time reduction in emergency rooms
   - Remote health control

4. *Agriculture:* Machine learning, along with IoT, can be used to increase the overall productivity of the crops and agriculture. It can help the farmers in automatic detection and diagnosis of plant diseases and automatic recommendations of required fertilizers.

TABLE 7.1

Various Applications of Machine Learning in IoT

Reference	IoT Application	Machine Learning Algorithm Used	Dataset	Performance
(Dogru and Subasi, 2018)	Traffic Accident Detection	Random Forest (RF)	Self-generated using SUMO	Accuracy: 92% Sensitivity: 94% Specificity: 88%
(Ng et al., 2019)	Road Surface Condition Identification	KNN, Random Forest, and SVM	IoT Sensors	Correctly Classified Rate: 99.43 (KNN), 99.14 (RF) 99.43 (SVM)
(Liang, 2015)	Automatic Traffic Accident Detection	SVM	Real World Traffic of Qingdao	Detection Rate: 99.85% False Alarm Rate: 1.58%
(Shakeel et al., 2018)	Security, Healthcare	Deep Q Learning Networks	IoT Sensors	Accuracy: 98.79%
(Azimi et al., 2018)	Healthcare	Deep Learning	MIT Arrhythmia (Moody et al., 2001)	Accuracy: 96%
(Goap et al., 2018)	Smart Irrigation	Machine Learning	Using IoT Sensors	Correlation Coefficient: 0.98 Accuracy: 0.96 Mean Square Error: 0.10
(Doshi et al., 2018)	Security against DDoS (Smart Homes)	• KNN • SVM • Decision Tree • RF	Using IoT Sensors	Accuracy: 99.9% (KNN) 99.1% (SVM) 99.9% (SVM, RF)

Table 7.1 provides an insight into various work in literature that has used the benefits of machine learning along with IoT.

7.6 Research Challenges and Future Directions in the Field of IoT Security Using Machine Learning and Deep Learning

Various research challenges in the use of machine learning and deep learning-based models for IoT security are presented in this section.

IoT and its related domains have huge potential, and machine learning has already proved to add to the numerous lists of benefits to the IoT network. However, to leverage the full benefits of machine learning in IoT, following major challenges need to be addressed:

1. *Pre-Processing of Data:* Data pre-processing involves the processing of the captured data using the IoT sensors for the removal of any type of ambiguous, redundant information, and for the filling of any missing information to get much from the gathered information. However, in IoT, data is gathered from a variety of sensors that are often constrained and discrete in nature and suffer from intermittent loss of connectivity. Because of this, the majority of the data contains irregularities and various

uncertainties, and missing and incomplete value is often found in the data, making the data pre-processing task bit complex in IoT. Proper pre-processing of data in IoT is a must for ensuring completeness of the data, which is crucial for ensuring high data quality. So, it is important to search for efficient pre-processing algorithms that can efficiently remove irregularities and uncertainties of the data, which are in turn important for further processing.

2. *Redundancy Reduction and Compression of Data:* Not all the data fetched from IoT Sensors are useful, and there exists a possibility of them containing redundant and similar data. There is a possibility of closely deployed IoT sensors to record similar nature of data which can impose redundancies in the captured data. This redundant data information can lead to various issues in the processing of the fetched data. It will not only lead to the wastage of computational energy required in the processing of this redundant information but will also lead to the wastage of storage which will be required in storing this redundant data. This redundant information will also lead to issues in the extraction of useful features which is crucial in optimal and efficient decision-making. So, it is required to have some efficient data redundancy removal techniques, which should be deployed to remove such similar information in order to ease the extra burden of storage and computation energy requirement.

3. *Still Prone to Adversarial and Other Attacks:* It has been observed and analyzed that almost all of the machine learning algorithms are prone to adversarial attacks, no matter which machine learning algorithm is being used (Zeadally and Tsikerdekis, 2020). The greater the inequality between an attacker and a defender, the more secure a machine learning model is from such attacks (Huang et al., 2011; Papernot et al., 2017). There are two attacks which are very easy to perform against the machine learning algorithms. One is the causative attack in which the attacker tries to tamper with the training or testing data to influence the model (Huang et al., 2011). Another attack is exploratory attack in which attacker tries to reverse engineer the machine learning-based security model to find out the model`s detection mechanism for making it able to avoid the detection (Shintre and Dhaliwal, 2019). So, more efforts are required to deal with these adverbial attacks. Similarly, another attack that can defeat machine learning model`s security is the poisoning attack in which an attacker can insert malicious data or script into the training data causing the wrong predictions, disastrous decisions making, and failure of the whole network which can also claim lives, as well as the privacy of people (Chen et al., 2018; Suciu et al., 2018). So, it is required to secure the machine learning models from such attacks and threats which will help in the overall strengthening of the IoT networking using the machine learning models.

4. *Protection against the Eavesdropping and Confidentiality Violation Attackers:* Eavesdropping attacks are somehow difficult to detect and identify, especially when the attacker is operating in the passive mode to overhear the communications traveling through various channels. There are, however, protection measures in the literature to deal with such confidentiality violation attacks like making use of cryptography and the use of encrypted channels. However, it is still possible for an attacker to extract sensitive information from the encrypted channels by spoofing the identities of some communication devices. So, it is required to explore some existing solutions for dealing with such passive and rogue spoofed entities by making use of some existing solutions along with the machine learning model.

7.7 Conclusion

IoT is a newly evolved technology that has seen a paradigm improvement in its technology over the years. IoT has a huge set of applications aiming to improve the overall living style of the people. Various research and applicability of IoT are focused on areas like agriculture, healthcare, intelligent transportation system, smart homes, etc. Due to some peculiar and inherent features of IoT, various security challenges are there in the deployment and implementation. So, to cope up with these security solutions, machine learning and deep learning algorithms and models can be applied, which can considerably improve the overall security of the network, ranging from existing attacks to zero-day attacks. In this study, an attempt has been made to explore the possibility and deployment challenges of machine learning-based security solutions for IoT.

References

Abbass, Wissam, Zineb Bakraouy, Amine Baïna, and Mostafa Bellafkih. 2018. "Classifying IoT security risks using deep learning algorithms," *2018 6th International Conference on Wireless Networks and Mobile Communications (WINCOM)*, Marrakesh, Morocco.

Al-Garadi, Mohammed Ali, Amr Mohamed, Abdulla Al-Ali, Xiaojiang Du, and J Mohsen. 2018. "A survey of machine and deep learning methods for internet of things (iot) security."

Azimi, Iman, Janne Takalo-Mattila, Arman Anzanpour, Amir M Rahmani, Juha-Pekka Soininen, and Pasi Liljeberg. 2018. "Empowering healthcare iot systems with hierarchical edge-based deep learning," *2018 IEEE/ACM International Conference on Connected Health: Applications, Systems and Engineering Technologies (CHASE)*, Washington D.C, USA.

BOC. 2014. Topologically Integrated Geographic Encoding and Referencing 2010. US %J United States Bureau of the Census. Available online: https://www.census.gov/geo/maps-data/data/tiger.html

Chen, Sen, Minhui Xue, Lingling Fan, Shuang Hao, Lihua Xu, and Haojin Zhu, Bo Li. 2018. "Automated poisoning attacks and defenses in malware detection systems: An adversarial machine learning approach," *Journal of Computers and Security*, 73:326–344.

Detrano, Robert, Andras Janosi, Walter Steinbrunn, Matthias Pfisterer, Johann-Jakob Schmid, Sarbjit Sandhu, Kern H Guppy, Stella Lee, and Victor J. 1989. "International application of a new probability algorithm for the diagnosis of coronary artery disease," *The American Journal of Cardiology Froeliche*, 64(5):304–310.

Dogru, Nejdet, and Abdulhamit Subasi. 2018. "Traffic accident detection using random forest classifier," *2018 15th Learning and Technology Conference (L&T)*, Jeddah, Saudi Arabia.

Doshi, Rohan, Noah Apthorpe, and Nick Feamster. 2018. "Machine learning DDos detection for consumer internet of things devices," *2018 IEEE Security and Privacy Workshops (SPW)*, San Francisco, CA, USA.

Goap, Amarendra, Deepak Sharma, AK Shukla, and C Rama Krishna. 2018. "An IoT based smart irrigation management system using Machine learning and open source technologies," *Journal of Computers and Electronics in Agriculture*, 155:41–49.

Guo, Cheng, Ruhan Zhuang, Chunhua Su, Charles Zhechao Liu, and Kim-Kwang Raymond. 2019. "Secure and efficient $\{K\}$ $ nearest neighbor query over encrypted uncertain data in Cloud-IoT ecosystem," *Journal of IEEE Internet of Things*, 6(6):9868–9879.

Ham, Hyo-Sik, Hwan-Hee Kim, Myung-Sup Kim, and Mi-Jung Choi. 2014. "Linear SVM-based android malware detection for reliable IoT services, *Journal of Applied Mathematics*, vol. 2014, Article ID 594501, https://doi.org/10.1155/2014/594501.

Huang, Ling, Anthony D Joseph, Blaine Nelson, Benjamin IP Rubinstein, and J Doug Tygar. 2011. "Adversarial machine learning," *Proceedings of the 4th ACM workshop on Security and artificial intelligence*, New York, United States.

Lakhina, Shilpa, Sini Joseph, and Bhupendra Verma. 2010. "Feature reduction using principal component analysis for effective anomaly–based intrusion detection on NSL-KDD."

Li, Jingfu. 2020. "IoT security analysis of BDT-SVM multi-classification algorithm," *Journal of International Computers, and Applications*, 1–10, DOI: 10.1080/1206212X.2020.1734313.

Liang, G.. 2015. "Automatic traffic accident detection based on the internet of things and support vector machine," *Journal of International Smart Home*, 9(4):97–106.

Moody, George B, and G. Roger. 2001. "The Impact of the MIT-BIH Arrhythmia Database," *Journal of IEEE Engineering in Medicine Mark, and Biology Magazine*, 20(3):45–50.

Moustafa, Nour, and Jill Slay. 2015. "UNSW-NB15: a comprehensive data set for network intrusion detection systems (UNSW-NB15 network data set)," *2015 Military Communications and Information Systems Conference (MilCIS)*, Canberra, ACT, Australia.

Ng, Jin Ren, Jan Shao Wong, Vik Tor Goh, Wen Jiun Yap, Timothy Tzen Vun Yap, and Hu Ng. 2019. "Identification of Road Surface Conditions Using IoT Sensors and Machine Learning." In *Computational Science and Technology*, 259–268. Springer.

Pajouh, Hamed Haddad, Reza Javidan, Raouf Khayami, Dehghantanha Ali, and Kim-Kwang Raymond. 2016. "A two-layer dimension reduction and two-tier classification model for anomaly-based intrusion detection in IoT backbone networks," *Journal of IEEE Transactions on Emerging Topics in Computing*, Vol. 7, pp: 314–323.

Papernot, Nicolas, Patrick McDaniel, Ian Goodfellow, Somesh Jha, Z Berkay Celik, and Ananthram Swami. 2017. "Practical black-box attacks against machine learning," *Proceedings of the 2017 ACM on Asia conference on computer and communications security*, Abu Dhabi United Arab Emirates.

Ring, Markus, Sarah Wunderlich, Dominik Grüdl, Dieter Landes, and Andreas Hotho. 2017. "Flow-based benchmark data sets for intrusion detection," *Proceedings of the 16th European Conference on Cyber Warfare and Security*, Dublin, Ireland. ACPI.

Sankaranarayanan, Suresh, and Srijanee Mookherji. 2019. "SVM-based traffic data classification for secured IoT-based road signaling system," *International Journal of Intelligent Information Technologies*, 15(1):22–50.

Shabtai, Asaf, Uri Kanonov, Yuval Elovici, Chanan Glezer, and Yael Weiss. 2012. "Andromaly": a behavioral malware detection framework for android devices," *Journal of Intelligent Information Systems*, 38(1):161–190.

Shakeel, P Mohamed, S Baskar, VR Sarma Dhulipala, Sukumar Mishra, and Mustafa Musa. 2018. "Maintaining security and privacy in health care system using learning based deep-Q-networks," *Journal of Medical Systems*, 42(10):186.

Sharma, Sparsh, and Ajay Kaul. 2018a. "Hybrid fuzzy multi-criteria decision making based multi cluster head dolphin swarm optimized IDS for VANET," *Journal of Vehicular Communications*, 12:23–238.

Sharma, Sparsh, and Ajay Kaul. 2018b. "A survey on Intrusion Detection Systems and Honeypot based proactive security mechanisms in VANETs and VANET Cloud," *Journal of Vehicular Communications*, 12:138–164.

Sharma, Surbhi, and Baijnath Kaushik. 2019. "A survey on internet of vehicles: Applications, security issues & solutions," *Journal of Vehicular Communications*, 20:100182.

Shen, Meng, Xiangyun Tang, Liehuang Zhu, Xiaojiang Du, and Mohsen Guizani. 2019. "Privacy-preserving support vector machine training over blockchain-based encrypted IoT data in smart cities," *IEEE Internet of Things Journal*, 6(5):7702–7712.

Shintre, Saurabh, and Jasjeet Dhaliwal. 2019. *Verifying that the influence of a user data point has been removed from a machine learning classifier*. Google Patents.

Suciu, Octavian, Radu Marginean, Yigitcan Kaya, Hal Daume III, and Tudor Dumitras. 2018. "When does machine learning {FAIL}? generalized transferability for evasion and poisoning attacks," *27th {USENIX} Security Symposium ({USENIX} Security 18)*, Berkeley, United States.

Tama, Bayu Adhi, and Kyung-Hyune Rhee. 2017. "Attack classification analysis of IoT network via deep learning approach," *Research Briefs on Information Communication Technology Evolution*, 3:1–9.

Tavallaee, Mahbod, Ebrahim Bagheri, Wei Lu, and Ali A Ghorbani. 2009. "A detailed analysis of the KDD CUP 99 data set," *2009 IEEE Symposium on Computational Intelligence for Security and Defense Applications*, Ottawa, Canada.

Vilela, Douglas WFL, T Ferreira Ed'Wilson, Ailton Akira Shinoda, Nelcileno V de Souza Araújo, Ruy de Oliveira, and Valtemir E Nascimento. 2014. "A dataset for evaluating intrusion detection systems in IEEE 802.11 wireless networks," *2014 IEEE Colombian Conference on Communications and Computing (COLCOM)*, Bogota, Colombia.

Zeadally, Sherali, and Michail Tsikerdekis. 2020. "Securing Internet of Things (IoT) with machine learning," *International Journal of Communication Systems*, 33(1):e4169.

8

Open-Source Tools for IoT Security

Anam Iqbal and Mohammad Ahsan Chishti

CONTENTS

8.1 Introduction to IoT Security Considerations

The Internet of Things (IoT) is an all-pervading aspect of our lives. It is no longer just a concept or 'the next big thing,' but a phenomenon surrounding us, a real technology-enabled, extensive network. The enabling technologies for IoT can transform any object into a smart object. Realizing IoT's full potential provides a framework that processes the raw input from things and transforms it into an intelligent output. An IoT environment revolves around increased communication between machines. These machines can be data-gathering sensors or actuators having embedded network connectivity, allowing these devices to be remotely controlled and send the collected data to higher-end devices for further analysis to derive meaningful insights (Dickson et al., 2018). However, given the heterogeneity and dynamic nature of an IoT ecosystem, a single enterprise cannot provide all the in-house expertise at every stage. Instead, they choose from an array of available integration partners or tools to simplify their work.

8.1.1 Security in IoT

The world in which we live today has accelerating technology, easy access to databases with data about vulnerable IoT devices, and sophisticated tools, all of which have led to a rapid increase in cyber threats and data leaks. Security has been a ripe area for IoT developers. The IoT environment shares similar security concerns as a mobile computing environment, but are in some cases more specific. Guaranteeing security necessitates protecting both IoT devices and services from unauthorized access from within and outside the network. It is pertinent to mention that there are many variations associated with the nature of nodes or objects in an IoT network like objects that can be continuously moving, or on the body, amplifies the security issues.

In less constrained devices, the security functionality overhead typically is expected to be 10–15% of the total system cost. These devices can support the security features and other operations so that all the entities in the environment are protected. However, when we shift the computing to constrained devices, the computing power also remains constrained (Greer et al., 2019). Security in this case follows a counter-intuitive cost model, as explained in Figure 8.1. In a constrained environment, the goal is to increase the device resource usage efficacy and decrease the vulnerability of the system. For this, the device needs to be made dominant in terms of defense, as nodes are directly in contact with the external environment, and contribute to 50% of the data leaks. The user has to configure these devices to perform security operations across all nodes and all levels. Hence the share of security-related functionality increases to as much as 60% of the total compute, leaving only 40% available for application goal-oriented computing. This increase results in the trinification of the application, thereby limiting the purpose of the built-embedded system (Geng, 2017).

We identified three fundamental problems with IoT devices and services: data confidentiality, privacy, and trust.

8.1.1.1 Data Confidentiality

In an IoT environment, data access is given to the user and the authorized objects. The IoT device, therefore, needs to verify if a user or another device is authorized to access a

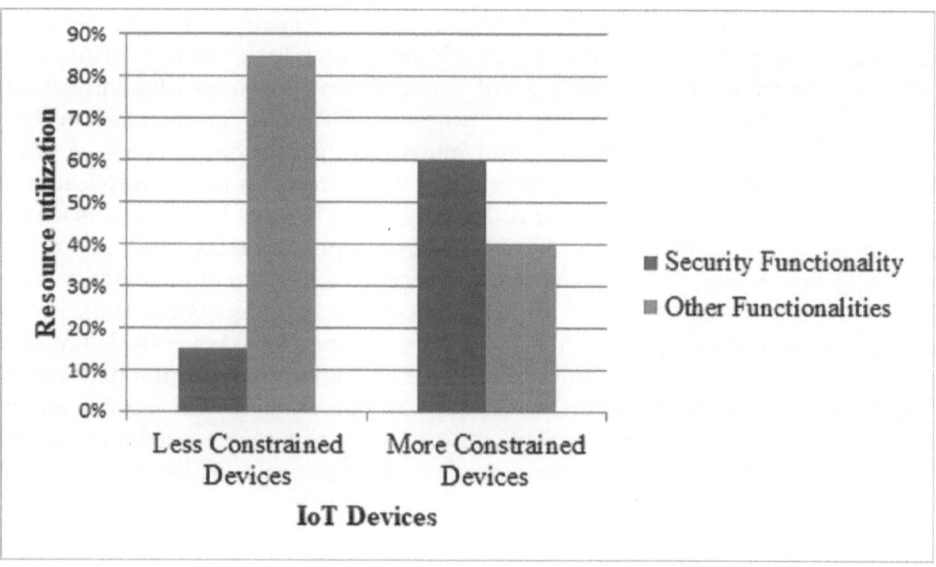

FIGURE 8.1
Cost model: security vs. other functionalities.

particular service, data associated, and the resources. The issue faced here is the creation, updating, and understanding of the access rules. Another aspect to be taken into consideration for data confidentiality is authentication. This issue is critical as multiple users and objects need to authenticate each other for better access.

8.1.1.2 Privacy

Two main elements are associated with privacy: data and communication networks. Data privacy is concerned with secure personal data collection and processing. The privacy of the communication network is concerned with protecting data sent over between two communicating devices.

8.1.1.3 Trust

An IoT network is immensely heterogeneous. Trust needs to be there in interactions between entities, as well as in the framework from the user perspective. The trustworthiness of an IoT device depends on the device components, both hardware, and software. For client trust, there needs to be an efficient mechanism for defining trust in a collaborative IoT environment (Atlam and Wills, 2020).

The security for 'things' in an IoT environment is fragmented and is somewhere the responsibility of the manufacturer. Nevertheless, there is a lack of a cohesive standard, which is a result of the absence of industry standards. The more significant task is to achieve a near real-time response from the security setup. Security implications in an IoT environment can arise at any step of the data path, during its acquisition, processing, or sharing from end nodes to the network layer and then to the application. One crucial factor when defending a potential threat is to cut down the rate at which the vulnerability impacts a system.

We can reduce this impact by implementing an efficient security architecture that covers all the layers of IoT architecture, including development and deployment, this involves not only securing the physical infrastructure. but also sanitizing the supporting infrastructure, the network, and smart devices connected. Physical security parameters include authorization and monitoring access to devices and hardware infrastructure. Network security includes the selection of suitable edge technology, right protocols for the application, and the suitable security configurations. For connecting smart devices to the servers and gateways, we need a system-level authorization. Security operations lay foundations on the threat intelligence collected over the network. The collection of threat intelligence is a vital issue since the devices connected to an IoT network are heterogeneous. Thus, data and device security are a perennial concern for every organization based over the IoT, with data security rooting its concerns from encrypted communication between the nodes and the network devices. While device security includes host operating system security and node security, all the OS or virtual environments must ensure proper user authentication, while the access permission mechanism implements node level security. While security concerns

TABLE 8.1

IoT Tools Based on IoT Layers

Layers	Elements	Tool
IoT End Points	Embedded operating system	Raspbian Ubuntu Core, RIOT
IoT local Network and IoT Internet	Frameworks software components and libraries	Node-RED IoTSys, OpenWRT, The ThingBox
Cloud & Data center	Cloud-based IoT platforms	SiteWhere, OpenRemote
Client devices	Device-side IoT platforms	OpenHab, Nimbits, IoT Toolkit

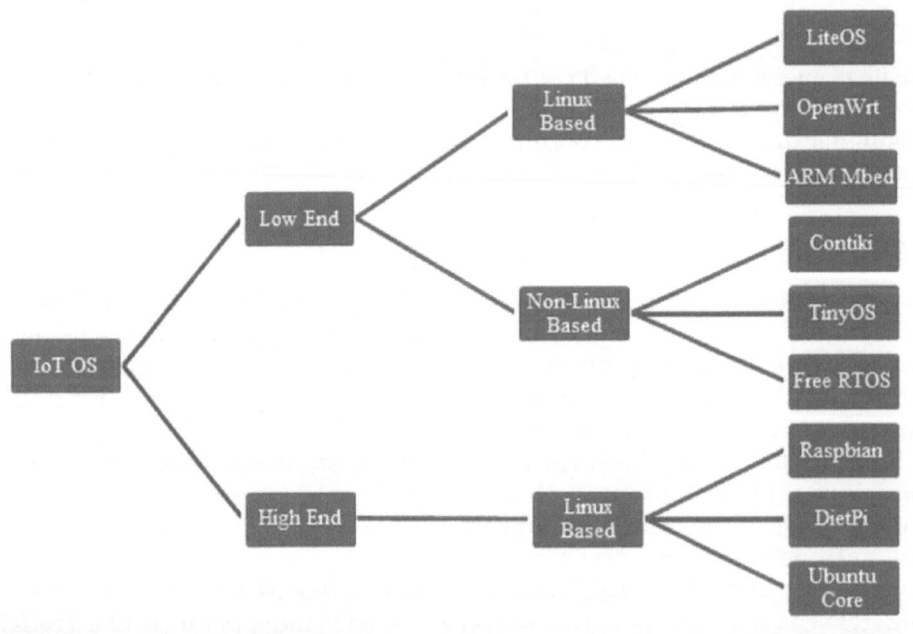

FIGURE 8.2
Types of IoT OSs.

each component of the IoT ecosystem—devices, platforms, tools, and applications—this chapter focuses on providing security using tools (Li, 2018; Nayyar, 2020).

In the list of parameters on deciding an IoT tool, security is one of the top priorities. Checking the features of an IoT tool before selecting it for an enterprise is essential. The least it should be able to provide is transport-level security for efficient communication between things, network layer, and the application. A few features that are taken into consideration when selecting a tool are the level of authorization, whether it provides data encryption if a security audit is available, and how complicated the lifecycle management of the tool is. Table 8.1 gives the list of IoT tools implementation on the different communication layers (Ojo et al., 2018; Cheruvu et al., 2019) (Figure 8.2).

8.2 Open-Source Operating Systems

A well-integrated IoT platform makes devices accessible to applications across enterprises. IoT subsystems are of paramount importance to industrial systems, in some cases having life safety implications (Zikria, 2018). However, connecting all our devices to the Internet, our data is exposed to high risks of vulnerability and data leaks, and this poses a severe challenge for IoT OSs to provide necessary security mechanisms, with minimal compromise on the functionalities, flexibility, scalability, and usability of the application. For choosing an operating system, certain limitations are to take into consideration, like processing power, RAM, storage memory. Figure 8.3 gives the general classification of IoT OS.

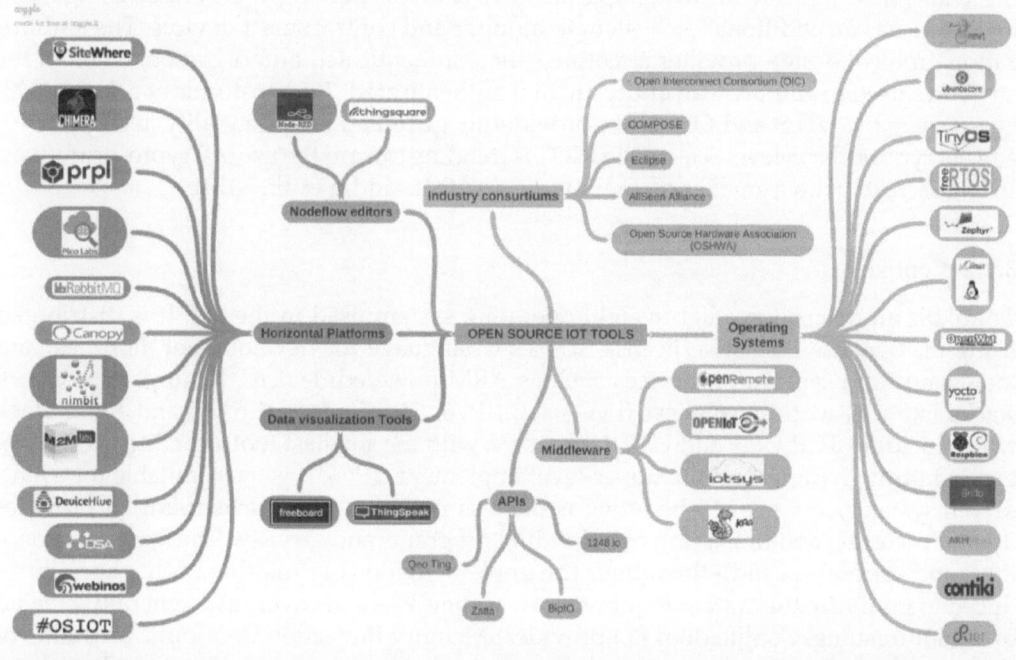

FIGURE 8.3
An infographic of IoT tools.

8.2.1 RIOT

RIOT is a free, open-source embedded operating system that supports most low-power devices. It supports mostly all microcontroller architecture, standard protocols, and network stacks such as UDP, TCP, CoAP, and IPv6. RIOT is a widely used operating system in developers because its development is easy and can be carried out under Linux using a native port, and it has a minimally hardware-dependent code. RIOT is a preferable choice by clients because of its robustness, multithreading support, modularity, code-footprint flexibility, energy efficacy, small memory footprint, uniform API access, and ultra-low interrupt latency (Anon., 2020a). The software modules of RIOT are stored separately and are aggregated only at compile-time, thus enabling modular completion of the system with only those modules being implemented that are required by the use case, thereby minimizing both system complexity and memory consumption. Tasks are executed in a multithreading environment and scheduled on a pre-emptive and tickle basis (Bansal and Kumar, 2020). A unique feature in RIOT is the power reservation technique available even in the sleep mode, in which memory is put into a low-power or 'not currently available' mode, which is a feature of the tickle scheduling, and wakes up the system only when an 'interrupt' occurs. RIOT fares better when it comes to memory usage and support. Applications of all sizes, ranging from smart homes and smart cities to wearables, have examples of prototypes and implementations using RIOT (Anon., 2019).

8.2.1.1 Security Services Provided by RIOT

A few modules or crypto-libraries that provide security support in RTOS are TweetNaCl high-security cryptographic library, qDSA (secure digital signatures), LibHydrogen cryptographic library (meant for constrained environments). One additional security feature is the availability of secure firmware upgrade, where before booting, it is verified using hashing. RIOT has an additional ecosystem to monitor and control smart devices. The authorization protocol design provides a feature called authenticated authorization, wherein the attributes used to authorize an entity are first authenticated. Two protocols at different OSI layers, namely, DTLS and OSCORE, provide integrity and confidentiality, using end-to-end encryption. Next level Security in RIOT is trending toward the use of Crypto-primitives, strong security with a smaller memory footprint. (Musaddiq et al., 2018).

8.2.2 Contiki

Contiki is an open-source lightweight operating system used in the IoT. It is distributed using the 3-clause BSD-style license. It uses C language for development, libraries, and tool-chains and can run on devices such as ARM-powered devices. It supports network stacks like uIP, world's smallest IPv6 stack (IPv6, 6TiSCH, and CoAP) and Rime Stack which includes TCP, UDP and HTTP protocols, with the smallest footprint. MAC is implemented through Rime, and the upper-level implements IPv6. It is most suitable for event-driven systems, and event-scheduling is done as per cooperative scheduling. It provides dynamic loading and unloading of individual programs and services. The kernel is monolithic and supports pseudo-threading. The implementation of protothreads has been introduced to minimize the system requirements, as they avoid the overhead generally, created by multithreading. Contiki does not provide a Memory Protection Unit, but a provision of changing the memory requirements during run time is available. Contiki provides persistent storage support through the Coffee Flash file system. It is a suitable choice for high-constrained devices due to its low memory and a low energy requirement. The simulator used with Contiki is Cooja (Musadiq et al., 2018; Bansal and Kumar, 2020).

8.2.2.1 Security Services Provided by Contiki

Contiki has ContikiSec (link-layer security) Transport Layer Security (TLS) which runs in three security modes, namely ContikiSec-Enc, for confidentiality only; ContikiSecAuth for integrity and authentication; ContikiSec-AE for confidentiality, integrity, and authentication. It also comes with cryptography libraries, and a fine-grained power tracing tool. Since the IoT devices are application-specific, security is an additional functionality and should be able to comply with a small memory footprint. ContikiSec is designed such that it maintains minimum energy consumption. One loophole in the ContikiSec is the absence of a defined key exchange mechanism, and instead, a public symmetric key, shared with all devices on the network, is used; this makes ContikiSec vulnerable to attack and data leaks from a compromised device (Anon., 2020b).

8.2.3 FreeRTOS

FreeRTOS is a freely distributed operating system under the MIT open-source license. It supports more than 40+ processor families like Cyclone V SoC, ARM Cortex-M33 simulator, and SAMV7. It is developed in C language. It includes a kernel and a huge set of libraries suitable for sorts of applications. FreeRTOS has a microkernel architecture. Tasks are executed in a multithreading environment. Since it is a real-time system, the memory allocation is dynamic, and scheduling is done in real-time on a cooperative and preemptive basis. As a power management technique, the system is sent into deep sleep mode when waiting for hardware or other resources. The FAT file system is provided for file management (Jaskani et al., 2019; Bansal and Kumar, 2020; Anon., 2020c).

8.2.3.1 Security Services Provided by FreeRTOS

Security support to RTOS uses WolfSSL. The main security requirements (e.g., authentication, integrity, confidentiality) are provided through WolfSSL (Zikria et al., 2019).

8.2.4 TinyOS

TinyOS is an open-source embedded operating system, initially developed by UC Berkeley in collaboration with Intel Corp, licensed under a New BSD license. TinyOS provides support to different hardware chips, from multi-controllers (Intel px27ax processor, Atmel family), to radio chips (variations of Infineon and Atmel) and flash chips (only the Atmel AT45DB and the STMicroelectronics STM25P chip). The networking support TinyOS is extensive and comprises of protocol designs that are parts of Internet standards. TinyOS supports Flooding Time Synchronization Protocol, Collection Tree Protocol, Drip, DIP, and DHV. Tiny OS is built upon a monolithic kernel and uses event-driven programming to perform non-preemptive tasks. TinyOS is written in the NesC language, a dialect of C. While waiting for other hardware components during scheduling, TinyOS implements power conservation using a split-phase and event-driven execution model. As long as there are no tasks in the task queue, the scheduler puts the CPU in sleep mode. A simulator used with TinyOS is TOSSIM (Anon., 2020d; Bansal and Kumar, 2020).

8.2.4.1 Security Services Provided by TinyOS

To provide security support to the TinyOS, it comes with a link-layer security solution called TinySec. TinyOS is better suited when resources are scarce, and every little bit of saved memory or computing power can help. Considering this, TinySec is designed to be as lightweight as possible, and its impact on the performance of the network is minimal. The three main security requirements—confidentiality, integrity, and authentication of an IoT network—are taken care of by TinySec. Additionally, TinySec implements Skipjack as the default block-cipher (Zikria et al., 2019).

8.3 Open-Source IoT Hardware Platforms

IoT hardware platforms form the backbone of any IoT ecosystem. It connects the devices to the cloud server through gateways. The parameters such as speed, the economy of scale, security and the complexity of the system are taken care of using the hardware platforms. It enables integration of the heterogeneous entities of the IoT environment (Hejazi et al., 2018).

8.3.1 DeviceHive

DeviceHive is a free, open-source machine-to-machine (M2M)/the hardware framework for communication between smart devices. It was released under Apache License 2.0. It comes with support for management protocols, multi-platform library, and data analysis functionality for the smart devices, by using RESTful API interfaces such as Swagger (Shilpa et al., 2019). It provides the functionality of monitoring the devices and controlling them using the device-management APIs for different protocols. MQTT and WebSocket protocols are provided as an additional plugin. All devices which support any of the three protocols can be connected to DeviceHive. The availability of monitoring tools and cloud-based API enables the user to start the discovery without being connected to the real hardware during the startup (Al-Taleb, 2019). The architecture of the DeviceHive is microservice based. With this architecture, it can connect to hundreds of devices simultaneously by generating the required number of instances to guarantee availability. DeviceHive owes its popularity to the library support it has, which includes Python, node.js, and java client libraries. It also comes with support for hardware sensors such as LM75A/LM75B/LM75C, using a special DeviceHive firmware (Anon., 2018). The examples of main areas of applications are smart home technology, real-time security, sensor-environments, smart energy, and telemetry (Anon., 2020e; Mehta, 2020).

8.3.1.1 Security Services Provided by DeviceHive

The authentication mechanism in DeviceHive is provided using JSON Web Tokens (JWT), where the tokens are stateless and self-contained; that is, they contain all the information about users, their privileges, details of connected devices and networks. The timeline for the token is customizable and has support for multiple back ends (Baccelli et al., 2013; Li, 2018).

8.3.2 Distributed Services Architecture (DSA)

DSA is an open-source hardware platform that facilitates communication at every layer of the IoT infrastructure. In DSA, the connection is established from the bottom to top, or upstream rather than downstream. Downstream places the request, and the upstream holds the authority to allow or deny the request. It integrates different devices and a variety of data protocols with one another in a decentralized manner. It represents each of the entities as one of these types: broker, DSLink, node, metric, data node, action attribute. By integrating these entities as one, tasks such as analytics, inter-device communication, and app development are simplified, as the load is distributed between discrete computing resources. The DSLinks run on edge devices and are connected to the brokers that manage the linking and permissions for nodes and security. Brokers are interconnected, thereby enabling an easy connection establishment between devices and cloud; this makes the tool a feasible choice for varied use–cases. The broker is upstream from the links. DSLinks also allow protocol translation from 3rd party data-sources. Based on the DSA is a popular platform called Cisco Kinetic EFM (Edge and Fog Processing Module). It provides hardware integration support using Dreamplug, Beaglebone, and Raspberry Pi (Anon., 2020f; User, 2020).

8.3.2.1 Security Services Provided by DSA

The connections in DSA being upstream are outbound and full-duplex transparently traversing firewalls and proxies. A special provision called quarantine is available in DSA. It is enabled on a broker; it can hold any unauthorized token in quarantine. The system can access the quarantined nodes, but they cannot access other nodes in the system. In DSA, all users are placed in different permission groups. If the 'superuser' value is enabled for any user, they don't need a permission group. The permission values represent the level of access allowed, some of the permission values being config, which is the highest level, followed by right (values can only be written), read (values can only be read), the list (only attributes can be read) and none. The permissions valid for a node are valid for all its child nodes (Li, 2018).

8.3.3 SiteWhere

SiteWhere is an industrial-strength open-source application training platform for the IoT that facilitates ingestion, storage, processing, and integration of device data. It provides a microserver-based infrastructure with the key resources needed to create and deploy IoT applications, and support high throughput and low latency. SiteWhere can run multiple applications on a single SiteWhere instance. It connects devices using MQTT, AMQP, SOMP protocols. It enables the addition of devices through self-registration (via REST services). One advantage is that it integrates with third-party development structures for high-level development. The default database store is MongoDB Eclipse Californium Framework for CoAP messages, InfluxDB for event data storage, Grafana to view SiteWhere data, and HBase for non-relational data storage (Anon., 2020g; Li, 2018).

8.3.3.1 Security Services Provided by SiteWhere

Spring is the main configuration framework that performs the bootstrapping, which also comes with a security infrastructure called Spring Security, with Apache Tomcat 7 and Hazelcast being the others (Nayyar, 2020).

8.4 Open-Source IoT Middleware

IoT middleware is an element of an IoT environment that provides a user with the raw data in a more representable and understandable format. The raw data is made available through actuators and sensors and needs to be presented in a format that is supported by all related applications. (Ngu et al., 2016) The use of an IoT middleware would avoid low-level reprogramming of the IoT environment. (Bansal and Kumar, 2020).

8.4.1 Kaa

Kaa is an end-to-end open-source middleware platform for business-level IoT that offers a wide range of modern Internet technologies. It has the flexibility to adapt it to any specific business domain, owning to its microservice architecture and endpoint SDK components. The endpoint SDK provides APIs for managing communication, data marshaling, and persistence (Anon., 2018). Kaa has been used extensively in the main sectors, such as industrial, health, telecommunications, smart solutions for cities, smart energy solutions, logistics, consumer electronics, and automobiles. The reason that this middleware has found its application in all walks of human life is that it facilitates cross-device interoperability, in addition to real-time device monitoring and configuration. The other reason for its popularity among developers is that it offers a wide range of options for implementing the network stack: Wi-Fi, Ethernet, ZigBee, MQTT, CoAP, XMPP, TCP, and HTTP (Anon., 2016; Anon., 2020h; Mehta, 2020).

8.4.1.1 Security Services Provided by Kaa

Kaa communication with objects is secured with TLS or DTLS, by default. For authentication, the rules are different for the server and the devices. The server is authenticated by the objects using the TLS certificate, while the devices are authenticated using pre-shared keys, or using client-side TLS certificates. The security infrastructure has the provision to allow the user to revoke device credentials. The device communication is secured with encryption and tamper protection, using certain microservices like Endpoint Lifecycle (EPL) services. The device management module of the platform provides support for device identity management that keeps a record of the endpoints, their access credentials, and any metadata related to them. The support for this module is provided by several microservice functionalities like endpoint register (ERP) and credential management (CM). The clients are authenticated using a username/password combination. The data that is collected from connected devices is sent either to a storage system or for further processing. The security of the data in the transmission is taken care of by EPL service. However, for reliable data storage, data collection microservice with the support of the data collection protocol enables the secure reception of data to endpoints and further to data receiver services for storage and processing (Anon., 2020h; Bansal and Kumar, 2020).

8.4.2 OpenIoT

OpenIoT is an open-source middleware initiative for integrating large scale IoT applications. It provides a means of data acquisition from sensor clouds. The actual sensors are abstracted, such that any other layer, including fog/edge, and it does not have to worry about the actual sensors. It provides a suitable interface between the things, and the cloud-based compute/storage or local compute/storage. It supports the majority of the standard protocols, MQTT, CoAP, IPv6, and HTTP. It derives its basis from the concept of sensing as a service. Open IoT has found a useful application in varied fields like smart agriculture for predicting crop yield and crowd-sensing (Katasonov et al., 2008; Fortino et al., 2018; Anon., 2020i).

8.4.3 OpenRemote

OpenRemote is an Open-Source Middleware for the IoT that allows any device to connect with the network, regardless of its type or the type of connection protocol. It proffers four types of integration tools, depending on the type of user. It supports several standards and protocols, including IPv6, oBIX, 6LoWPAN, and Constrained Application Protocol. It aims at subduing the challenges faced during the integration of devices. It also includes a rules engine that we can program with a workflow- or flow-based editor, and UI components for front-end developers (Anon., 2019; Harwood, 2019; Anon., 2020j).

8.5 Open-Source APIs

Open-source APIs provides a platform for data acquisition, data processing, data presentation and node management (Anon., 2019).

8.5.1 Zetta

Zetta is an open-source, API oriented IoT platform. It is built on Node.js. It is regarded as a complete toolkit to generate HTTP APIs for devices and IoT servers that are either on the cloud or at geographically distant locations. It combines REST APIs, WebSockets to build real-time applications. The IoT platform's architecture comprises the following components: server, scouts, drivers, server extensions, and registry, with the server being the top one on the list. It can convert any device into an API (Sakovich, 2017; Li, 2018; Anon., 2020k).

8.5.2 BipIO

BipIO is an HTTP API that is used in easy visualization to integrate Web-based APIs with our devices. It implements graph-based pipelines (bips) on the endpoints. It is a RESTful JSON API. One of the most significant advantages it has is that it can be reconfigured dynamically, without having to change the implementation of the client (Mineraud et al., 2016; Anon., 2020l).

8.6 Discussion

We surveyed and compared mainstream open-source tools on the technical support they offer in an IoT environment and is illustrated in Table 8.2. Analyzing the comparative can give us an insight as to which tool is more appropriate for any particular application.

TABLE 8.2

Functionalities of Various Open-source IoT Tools

IoT Tool	Device Management	Data Collection	Network Stack	Security	Integration	Analytics	Storage	Visualization
Contiki	✓	REST API	uIP, RIME	ContikiSec	REST API	✓	×	×
RIOT	×	CoAP, MQTT	GNRC, CCNlite	Crypto-libraries	REST API	×	×	×
FreeRTOS	✓	MQTT, HTTP	None, but supports third-party stacks	WolfSSL	REST API	✓	✓	×
TinyOS	✓	CTP	TYMO	TinySec	REST API	✓	✓	✓
ARM mbed	✓	REST API, MQTT	LwIP TCP/IP	SSL/TLS, X.509 Certificate	REST API	×	×	×
Ubuntu Core	✓	MQTT, AMQP	—	AppArmor, Seccomp	REST API	×	✓	×
DeviceHive	✓	REST API, MQTT	WebSocket, TCP/IP	JWT	REST API, MQTT	✓	✓	✓
SiteWhere	✓	MQTT, JSON, AMQP, WebSockets	gRPC	Spring Security, SSL	REST API, gRPC	✓	✓	×
IoTivity	✓	CoAP, Message Queue	TCP/IP	DTLS/TLS	REST API	✓	×	×
DSA	×	HTTP	UDP	Basic authentication by broker	REST API	✓	✓	×
Cylon.js	✓	REST API, MQTT	TCP/IP	×	REST API	×	✓	×
Kaa	✓	MQTT, CoAP	TCP/IP	TLS/DTLS, RSA, AES, and EPL Services.	REST API	✓	✓	✓
Netbeast	✓	HTTP, MQTT	UDP, TCP/IP	TLS/SSL	REST API	✓	✓	✓
ThingsBoard	✓	MQTT, CoAP, HTTP	TCP/IP, UDP	SSL, access tokens	REST API	✓	✓	✓
Brillo	✓	REST API	TCP/IP	SSL/TLS	REST API	✓	✓	✓
OpenIoT	×	MQTT, CoAP	TCP/IP	User Management, authentication and authorization	REST API	✓	✓	✓

Except for RIOT, Nimbits, DSA, and Open IoT, the rest of the tools provide device management functionality. RIOT, Contiki, TinyOS, Brillo, do not support the native security feature but instead allow the third-party plugins to take care of their security needs (Anon., 2017). In most of the tools, MQTT, CoAP, and REST APIs provide data collection, except for DSA and TinyOS, both using language-specific scripts for device management. Analytics, storage, and visualization functionalities by tools are of paramount importance, considering the variety and volume of data IoT devices generate daily (Franklin, 2020). RIOT, ARM mbed, Ubuntu Core, and Cylon.js platforms do not have analytics support.

Raw data is gathered using sensors and then fed to the rest of the architecture, and this makes these tools unsuitable for data-critical applications, like smart healthcare. Other than Cylon.js, all other tools discussed have embedded security mechanisms to tackle any security breach in terms of authentication, data integrity, and confidentiality. Additionally, some of these platforms support not only remote device configuration and control but also over-the-air firmware updates (Solangi et al., 2018).

8.7 Conclusion

The chapter illustrates the parameters for choosing an IoT cloud solution based on the use of open-source tools. We discussed the device management, network stack, resource management, and security features of Contiki, TinyOS, RIOT, FreeRTOS, DeviceHive, DSA, SiteWhere, Kaa, OpenIoT, OPenRemote, Zetta and BipIO. We have also highlighted their advantages, disadvantages, and applications. The dramatic proliferation of the IoT has made it customary to share our details with the smart devices which are connected to the Internet all the time, and stand at a high risk of security breach and data leaks over the Internet. In this chapter, we have analyzed various open-source tools. We have provided an overview of the multiple functionalities of these tools, with our focus being security. All tools at ground level have been compared based on capabilities like device support, computation (analysis and visualization), communication interfaces, and security support mechanism. From the content of this chapter, we can conclude that, although all the discussed platforms are very well known for IoT solution development, a few OS stand out. TinyOS is best in terms of resource conservation; event-driven Contiki is most prevalent in dedicated environments since power consumption and memory footprint are a priority, while RIOT is the most prominent in multithreaded systems.

Implementing these tools in some real-world IoT environments can further elaborate on the advantages and disadvantages each of these tools has since, in real-time, there are limitations, such as energy consumption, validation, and calibration, which we cannot simulate.

References

Al-Taleb, N., and Min-Allah, N., February 2019. A Study on Internet of Things Operating Systems. *IEEE International Conference on Electrical, Computer and Communication Technologies (ICECCT),* 1–7. IEEE. doi: 10.1109/ICECCT.2019.8869062

Anon. 2016. Plataformas IoT. *Capterra*. Retrieved March 21, 2020, from https://www.capterra.es/directory/31016/iot/software.

Anon. 2017. Evaluating Open-Source IoT Platform Vs. Serverless IoT Platform. *IoT for All*. Retrieved March 21, 2020, from https://www.iotforall.com/iot-platform-open-source-vs-serverless/.

Anon. 2018. IoT Tools|concept and a few popular IoT development rools. *EDUCBA*. Retrieved March 19, 2020, from https://www.educba.com/iot-tools/.

Anon. 2019. 10 Open-Source Tools For The Internet Of Things. *Avsystem.Com*. Retrieved March 23, 2020 from https://www.avsystem.com/blog/10-open-source-iot-tools/.

Anon. 2020a. RIOT - The Friendly Operating System For The Internet Of Things. *Riot-Os.Org*. Retrieved March 16, 2020, from https://www.riot-os.org/.

Anon. 2020b. Contiki-Ng/Contiki-Ng. *Github*. Retrieved March 20, 2020 from https://github.com/contiki-ng/contiki-ng/wiki.

Anon. 2020c. Freertos - Market Leading RTOS (Real-Time Operating System) For Embedded Systems With Internet Of Things Extensions. *Freertos*. Retrieved March 15, 2020, from https://www.freertos.org/.

Anon. 2020d. TinyOS Home Page. *Tinyos.Net*. Retrieved March 19, 2020, from http://www.tinyos.net/.

Anon. 2020e. Devicehive - Open Source IoT data platform with the wide range of integration options. *Devicehive.Com*. Retrieved March 22, 2020, from https://devicehive.com/.

Anon. 2020f. Open Source IoT Platform & Toolkit | DSA - Home. *Iot-Dsa.Org*. Retrieved March 23, 2020, from http://iot-dsa.org/.

Anon. 2020g. Sitewhere Open Source Internet Of Things Platform. *Sitewhere.Io*. Retrieved March 17, 2020, from https://sitewhere.io/en/.

Anon. 2020h. Internet Of Things Platform | Kaa. *Kaa IoT Platform*. Retrieved March 21, 2020, from https://www.kaaproject.org/.

Anon. 2020i. Openiot – Open Source Cloud Solution For The Internet Of Things. *Openiot.Eu*. Retrieved March 21, 2020, from http://www.openiot.eu/.

Anon. 2020j. Openremote | Open Source For Internet Of Things. *Openremote.Com*. Retrieved March 21, 2020, from http://www.openremote.com/.

Anon. 2020k. Zettajs/Zetta. *Github*. Retrieved March 21, 2020, from https://github.com/zettajs/zetta.

Anon. 2020l. Bipio-Server/Bipio. *Github*, Retrieved March 21 2020, from https://github.com/bipio-server/bipio/wiki.

Atlam, H.F. and G.B. Wills. 2020. IoT Security, privacy, safety and ethics. In *Digital Twin Technologies and Smart Cities* (pp. 123–149). Springer, Cham. doi: 10.1007/978-3-030-18732-3_8

Baccelli, E., O. Hahm, M. Gunes, M. Wahlisch and T.C. Schmidt. April 2013. RIOT OS: Towards an OS for the Internet of Things. *IEEE Conference on Computer Communications Workshops (INFOCOM WKSHPS)*, 79–80. IEEE. doi: 10.1109/INFCOMW.2013.6970748

Bansal, S., and D. Kumar. 2020. IoT ecosystem: a survey on devices, gateways, operating systems, middleware and communication. *International Journal Of Wireless Information Networks*. Springer Science and Business Media LLC. doi: 10.1007/s10776-020-00483-7

Cheruvu, Sunil, Anil Kumar, Ned Smith, and David M. Wheeler. 2019. *Demystifying Internet of Things Security: Successful IoT Device/Edge and Platform Security Deployment*. Apress. doi: 10.1007/978-1-4842-2896-8_1

Dickson, R., H. Lisachuk, A. Ogura, and M. Cotteleer. 2018. Growing Internet of Things Platforms. Deloitte Insights. Retrieved 17 March 2020, from https://www2.deloitte.com/us/en/insights/focus/internet-of-things/iot-ecosystem-platforms-value-creation.html.

Greer, C., M. Burns, D. Wollman and E. Griffor. 2019. Cyber-physical systems and Internet of Things. *NIST Special Publication*, 1900, 202. doi: 10.6028/NIST.SP1900-202

Fortino, Giancarlo, Claudio Savaglio, Carlos E. Palau, Jara Suarez de Puga, Maria Ganzha, Marcin Paprzycki, Miguel Montesinos, Antonio Liotta, and Miguel Llop. 2018. Towards multi-layer interoperability of heterogeneous IoT platforms: The INTER-IoT approach. *Integration, Interconnection, and Interoperability of IoT Systems* (pp. 199–232). Springer, Cham. doi: 10.1007/978-3-319-61300-0_10

Franklin Jr, C. 2020. 7 Tools For Stronger IoT Security, Visibility. *Dark Reading*. Retrieved March 21, 2020, from https://www.darkreading.com/endpoint/7-tools-for-stronger-iot-security-visibility/d/d-id/1331824?image_number=8.

Geng, Hwaiyu. 2017. Internet of Things and data analytics in the cloud with innovation and sustainability. In *The Internet of Things & Data Analytics Handbook* (pp. 3–28). doi: 10.1002/9781119173601.ch1

Harwood, T. 2019. IoT Software | 2019 Guidebook On Tools, OS And Frameworks. *Postscapes*. Retrieved March 21, 2020, from https://www.postscapes.com/internet-of-things-software-guide/#open.

Hejazi, Hamdan, Husam Rajab, Tibor Cinkler, and László Lengyel. 2018. Survey of platforms for massive IoT. *IEEE International Conference on Future IoT Technologies (Future IoT)*, 1–8. IEEE. doi: 10.1109/FIOT.2018.8325598

Jaskani, F.H., S. Manzoor, M.T. Amin, M. Asif and M. Irfan. 2019. An investigation on several operating systems for internet of things. *EAI Endorsed Transactions on Creative Technologies*, 6(18). doi: 10.4108/eai.13-7-2018.160386

Katasonov, A., O. Kaykova, O. Khriyenko, S. Nikitin and V.Y. Terziyan. 2008. Smart semantic middleware for the Internet of Things. *Icinco-Icso*, 8: 169–178. 10.1016/j.comcom.2016.03.015

Li, YangQun. 2018. An integrated platform for the internet of things based on an open source ecosystem. *Future Internet*, 10(11): 105.doi: 10.3390/fi10110105

Mehta, N. 2020. 10 most popular open source IoT frameworks. *Techtic Solutions*. Retrieved March 23, 2020, from https://www.techtic.com/blog/top-10-open-source-iot-frameworks/

Mineraud, J., O. Mazhelis, X. Su and S. Tarkoma. 2016. A gap analysis of Internet-of-Things platforms. *Computer Communications*, 89: 5–16. doi: 10.1016/j.comcom.2016.03.015

Musaddiq, A., Y.B. Zikria, O. Hahm, H. Yu, A.K. Bashir and S.W. Kim. 2018. A survey on resource management in IoT operating systems. *IEEE Access*, 6: 8459–8482. doi: 10.1109/ACCESS.2018.2808324

Nayyar, A. 2020. *Open Source For You*. Retrieved March 17, 2020, from https://opensourceforu.com/author/anand-nayyar/.

Ngu, Anne H., Mario Gutierrez, Vangelis Metsis, Surya Nepal, and Quan Z. Sheng. 2016. "IoT middleware: A survey on issues and enabling technologies." *IEEE Internet of Things Journal*, 4(1): 1–20. doi: 10.1109/JIOT.2016.2615180

Ojo, Mike O., Stefano Giordano, Gregorio Procissi, and Ilias N. Seitanidis. 2018. A Review of low-end, middle-end, and high-end iot devices. *IEEE Access*, 6: 70528–70554.doi: 10.1109/ACCESS.2018.2879615

Sakovich, N. 2017. The Most Popular IoT Development Tools and Technologies | Sam Solutions. Retrieved March 21, 2020, from https://www.sam-solutions.com/blog/the-best-tools-for-internet-of-things-iot-development.

Shilpa, V., Vidya, A., Chandrashekhara S.N., 2019. Learning On Tools Used In IoT Development Life Cycle. *International Journal Of Innovative Technology And Exploring Engineering* 9(2S): 701–707. Blue Eyes Intelligence Engineering and Sciences Engineering and Sciences Publication - BEIESP. doi: 10.35940/ijitee.B1019.1292S19

Solangi, Z.A., Y.A. Solangi, S. Chandio, M.S. Bin Hamzah and A. Shah. May 2018. The future of data privacy and security concerns in Internet of Things. *IEEE International Conference on Innovative Research and Development (ICIRD)*, 1–4. IEEE. doi: 10.1109/ICIRD.2018.8376320

User, S. 2020. Open Source IoT Platform & Toolkit | DSA - How DSA Works. Iot-Dsa.Org. Retrieved March 20, 2020, from http://iot-dsa.org/get-started/how-dsa-works.

Zikria, Y., S. Kim, O. Hahm, M. Afzal, and M. Aalsalem. 2019. Internet Of Things (IoT) operating systems management: opportunities, challenges, and solution. *Sensors*, 19(8): 1793. doi: 10.3390/s19081793

Zikria, Y., H. Yu, M. Afzal, M. Rehmani, and O. Hahm. 2018. Internet Of Things (IoT): operating system, applications and protocols design, and validation techniques. *Future Generation Computer Systems*, 88: 699–706. doi: 10.1016/j.future.2018.07.058

9

A Boolean Approach for Computational Design of Ethics

Sahil Sholla

CONTENTS

9.1 Introduction

IoT is a paradigm shift in our understanding of networking that endeavors to connect even ordinary things to the internet. 'Ordinary things' are embedded with sensing, communication and processing abilities. The data collected by smart devices is processed in order to provide advanced class of services. IoT offers a huge applications market like smart healthcare, smart transportation systems, smart grid and so on. Given the vast range of applications IoT offers, it has the potential to transform our society.

Notwithstanding the benefits IoT promises, ever-increasing presence of smart things that are able to collect huge amount of personal and social information raises ethical concerns. This is particularly true when smart devices are endowed with artificial intelligence capability to execute decisions autonomously. Ethics symbolize the collection of beliefs, values and morals of a given society that serve as guiding principles for amicable mutual coexistence. Ethics, morality and culture serve as essential vehicles for smooth functioning of any society.

As devices become more pervasive in our personal and social lives, they need to function in an ethically consistent manner. Autonomous decision-making on the part of machines that involve arbitrary ethical decisions may disrupt social order and lead to technological chaos. While a smart device is a logicaldecision-making device, it does not mean autonomous decisions taken by smart devices should be disconnected from human agency. Smart devices do not operate in isolation; rather decisions they take have a bearing on entities external to the device as well. The way smart devices operate and execute autonomous decisions have a direct influence on our moral and social lives, and thus, such factors should be taken into consideration (Jasanoff, 2016). Technological artifacts essentially lack moral agency and are not morally accountable for themselves. Therefore, to allow decision-making independent from any external factors (even if that was possible) is tantamount to intellectual suicide. Also, human biases are inevitably reflected in the design of any technological artifact including computer software and mathematical models (Friedman et al., 2006). A sound set of social, cultural and ethical norms represent essential components of a healthy society that need to be taken into consideration when the technologies are implemented. Thus, ethical connotations of technology cannot be undermined else wellbeing of the very people IoT seeks to benefit could be endangered. A detailed discussion on the ethical implications of IoT can be found in Sholla et al. (2018). In this work, we outline a method that can be used to implement context specific ethics in smart devices to facilitate the development of social friendly IoT.

9.2 Methodology

Usually the workings of a smart device only focus on technical specifications for efficient design. Devices are treated as stand-alone systems disconnected from moral, ethical and legal context in which they operate. However, this may inadvertently result in situations that compromise essential human values or cause serious physical injury, even loss of life. The problem is exacerbated in the face of artificial intelligence enabled pervasive devices of the IoT world. Thus, it is necessary to consider the question of ethical complicity of smart devices in order to safeguard human interests.

The question of endowing machines with the capability of moral decision-making is studied in an area of learning called machine morality, also known by the terms roboethics, machine ethics, or friendly AI (Artificial Intelligence) (Wallach and Allen, 2009). The question is difficult because religious scholars, ethicists and philosophers differ on the topic. Also ethical choices vary among people based on society age etc. The interdisciplinary nature of the question that includes psychology, sociology, philosophy, religion and AI makes it even more challenging (Anderson and Anderson, 2011). Machine morality aims to design Artificial Moral Agents (AMAs) that are able to exhibit ethically sound behavior based on ethical, moral, philosophical or religious grounds (Allen et al., 2005). However, such full-fledged moral judgment capability may not necessary for the smart things of the IoT.

One of the earliest scientific approaches to design morality wooden machines is the Software System called JEREMY (Anderson et al., 2004). The software utilizes utilitarian ethics in order to evaluate significance of various actions. It uses intensity duration and probability has the three parameters decide on the ethical nature of actions. A program requires a user to enter each action's description and a measure of corresponding pleasure and likelihood of its occurrence. This information is used by the program to calculate net pleasure associated with each action, and the program chooses the action that achieves maximum pleasure as the future course of action. Another attempt to design ethical sensitivity within devices is a program called MedEthEx (Anderson et al., 2006). It uses the philosophy of prima facie duties to resolve ethical dilemmas in biomedical context. Values corresponding to three prima facie duties of beneficence, non-maleficence and respect are used to guide a medical professional to choose among the given possible actions that could be taken in a given scenario. However, the design approach pertains to a specific example of a patient refusing to take medicine for a severe health condition and is not amenable to extension in other domains. In a recent work Baldini et al. (2016) use a policy-based approach to denote ethical choices onsmart devices. However,the authors view ethics in a limited perspective, focusing on privacy profiles for each device. The end users have the option to indicate their ethical preferences using these privacy profiles for which they also need to pay the service provider.

We propose to incorporate context specific ethical requirements of smart devices within their design. Devices would operate within a particular moral ethical context that would only necessitate a subset of ethical decisions. It is plausible to focus on these essential ethical requirements rather than a method toward universal moral theory. We adopt an ontology agnostic engineering approach to ethics implementation. We do not concern ourselves with the philosophical debate surrounding ontology of ethics. Our objective is to implement a given set of moral, ethical or legal considerations applicable in the context of a smart device regardless of the philosophy that motivates such considerations.

Human beings, regardless of religion, ethnicity, race, region and society, share a rich reservoir of ethics that appears promising for inclusion within device functioning. We refer to the collection of sociological, ethical, legal or religious parameters relevant to the context of a particular machine as collectively as Ethics of Operation (EOP). The proposed method works as follows: First we express the EOP in the form of propositional statements (whether related to input or output). With the help of propositional variables, we identify various scenarios that may result between the smart device and its context. The scenarios could represent normal functional behavior, malfunction, ethical behavior, as well as ethical breaches. Various combinations of the propositional variables and the meaning implied by them is indicated in the context table. Then, EOP is used to design appropriate ethical response. For each scenario in the context table corresponding ethical response is expressed in the manners table. The mapping from the context table to manners table enables the machine to exhibit ethical behavior. Such incorporation ensures essential human values that invariably come into play in the context of smart device functioning are upheld.

9.3 The Proposed Approach

The ethical requirements of the device based on context requirements are analyzed to clearly identify the device EOP. The EOP is expressed in terms of propositional statements p^1, p^2, p^3... p^n. A proposition is a declarative sentence (i.e., a sentence that declares a fact that can either have a true or false value). It is referred to as propositional because it asserts a truth value or makes a proposition. For instance, 'Delhi is the capital of India' is a true proposition, whereas, '2+3=6' is a false preposition. The sentence, 'What day is it?' is not a propositional statement because it is not declarative. Similarly, 'x>2' is not a propositional statement because the variable x is unknown. However, if a specific value is assigned to x, it becomes a proposition. A detailed account of propositional statements and related concepts can be found in Rosen (2009). Truth values of the propositional statements represent various scenarios that occur while the machine functions in its context. These scenarios may represent normal device functioning, erroneous working of the device (that may or may not have ethical implications), ethical machine behavior or a conflict with device EOP (i.e., ethical breaches).

In order to facilitate understanding of the various scenarios and their ethical nature, we generate a table fusing the propositional variables called context table. The context table

TABLE 9.1

Context Table

S.No	p^1	p^2	p^3	$\cdots$	p^n	Ethical Status
1	0	0	0	$\cdots$	0	e_1
2	0	0	0	$\cdots$	1	e_2
$\vdots$	$\vdots$	$\vdots$	$\vdots$	$\vdots$	$\vdots$	$\vdots$
k	0	0	1	$\cdots$	0	e_k
$\vdots$	$\vdots$	$\vdots$	$\vdots$	$\vdots$	$\vdots$	$\vdots$
2^n	1	1	1	$\cdots$	1	e_{2^n}

indicates all possible combinations of truth values that may result from the propositions. Structure of a typical context table is indicated in Table 9.1.

Where e_k denotes ethical status of scenario k (e.g., forbidden, permissible or obligatory). We use these three ethical status types to indicate ethical desirability of different possible scenarios with an application context. Forbidden scenarios represent those scenarios that are inconsistent as per machine EOP and thus are prohibited. Whereas obligatory scenarios are essential ethical requirements that a machine needs to fulfill. Permissible scenarios may or may not be allowed for the machine based on the ethical expectations of machine functioning.

Based on the EOP, appropriate ethical response for each scenario is indicated in another table called the manners table. The manners table contains only ethically consistent truth values for each scenario. Truth values of the propositions corresponding to ethically compatible scenarios are kept as such. However, truth values of ethically inconsistent scenarios (forbidden and/or permissible) are modified (inverted) to ethically acceptable ones. The mapping of scenarios in the context table to ethically consistent scenarios in the manners table is shown in the manners map as shown in Table 9.2. This mapping enables a machine to exhibit ethical behavior in its context as per EOP.

The ethically consistent truth values for propositional variable in the manners table may also be described using Boolean equations. For each column of the manners table, we focus on the rows that are set to '1' and consider corresponding rows in the context table to evaluate the Boolean equation.

TABLE 9.2

Manners Map

S.No	p1	p2	p3	$\cdots$	pn
1	0	0	0	$\cdots$	0
2	0	0	0	$\cdots$	1
3	0	0	1	$\cdots$	0
$\vdots$	$\vdots$	$\vdots$	$\vdots$	$\vdots$	$\vdots$
2^n	1	1	1	$\cdots$	1

$\Longrightarrow$

P1	P2	P3	$\cdots$	Pn
0	0	1	$\cdots$	1
0	0	0	$\cdots$	1
1	0	1	$\cdots$	0
$\vdots$	$\vdots$	$\vdots$	$\vdots$	$\vdots$
0	0	0	$\cdots$	1

Algorithm used for implementing ethics in the healthcare device and its complexity analysis is as follows:

Algorithm Step	Complexity
1. retrieve variables: health emergency status (e), authorization (a) and message content (s)	$O(1)$
2. enumerate 2^n possible scenarios, where n is the number of variables	$O(2^n)$
3. set/unset bits in manners map to design ethical response to scenarios	$O(2^n)$

Complexity of the algorithm is $O(2^n)$. As the number of devices increase exponentially and the situation that a particular IoT device has to encounter increases in complexity the number of proportional variables that sufficiently capture the ethical requirements of context also increase exponentially. This in turn would require IoT devices to handle an exponentiallyincreasing number of possible scenarios. However, focusing on context specific requirements tailored to a particular application context fends off the problem of scalability to some extent. Also, as IoT devices continue to evolve in terms of processing power and memory capability, a greaternumber of proportion variables can be handled.

9.4 Use Case

To demonstrate the working of the proposed method implementing ethics, we consider a healthcare device with ethical requirements as specified in Table 9.3. We denote the Ethics of Operation (EOP) in the smart healthcare context using Boolean variables l, m, n that indicate:

l: health condition

m: authorized to view

n: emergency SOS

We construct a matrix to represent the context table using these propositional variables to reflect various scenarios that can occur in the interaction between the smart healthcare product and its environment. These propositions take either true (T) or false (F) values that we denote by 1 and 0, respectively.

TABLE 9.3

Context Table for the Healthcare Device

Scenario	l	m	n	Meaning	Ethical Status
1	0	0	0	no emergency, noauthorization, sos false	obligatory
2	0	0	1	no emergency, no authorization, sos true	forbidden
3	0	1	0	no emergency, authorization, sos false	obligatory
4	0	1	1	no emergency, authorization, sos true	forbidden
5	1	0	0	emergency, no authorization, sos false	forbidden
6	1	0	1	emergency, no authorization, sos true	permissible
7	1	1	0	emergency, authorization, sos false	forbidden
8	1	1	1	emergency, authorization, sos true	obligatory

TABLE 9.4

Manners Map for the Healthcare Device

Scenario	Context Table			Manners Table		
	l	m	n	L	M	N
1	0	0	0	0	0	0
2	0	0	1	0	0	0
3	0	1	0	0	1	0
4	0	1	1	0	1	0
5	1	0	0	1	0	1
6	1	0	1	1	0	1
7	1	1	0	1	1	1
8	1	1	1	1	1	1

The use case considers three variables for ethical conformity of a smart device. Thus, sample space is $2^3 = 8$. Each scenario is the sample space is associated with some ethical status. For instance, Scenario 1 considers a situation where the device does not detect any emergency medical condition and a particular recipient who is not authorized to view health related information of the patient. In such care, it is obligatory to not disclose any health related information to such individual. In Scenario 7, there is a medical emergency and recipient is also authorized. In such circumstances, not sending the health-related information (SOS) to recipient is forbidden for smart device. Conversely, in Scenario 8, all three variables are set to 1, and thus such behavior is obligatory for the smart device. In a similar fashion other scenarios can be explained. The meaning of different codes and their ethical status is shown in Table 9.3.

We define Boolean variables L, M and N for variables l, m and n respectively, that provide ethics compliant values to ensure that the healthcare product performs according to the ethical policies adopted. Next, we construct the manners map, a mapping between various scenarios in the context table to corresponding ethically compliant scenarios in the manners table as shown in Table 9.4.

In the above table the possible EOP conflicts have been handled as per ethics policies (shown in **bold** in the manners map). In order to model different ethical scenarios that may occur during the healthcare device functioning, we can consider 100 runs of the simulation, each triggering a scenario from the context table randomly.

9.5 Discussion

Since the idea of smart city encompasses multiple areas of societal welfare, interdisciplinary work would be necessary to materialize it. Ethical requirements from the government, law enforcement, industry and people using the technology could be combined carefully by establishing precedence and removing ambiguities to form unified device EOP.

The proposed method of implementing ethics needs human assistance to design the manners map from a given EOP. For smaller examples this could be easily achieved, as illustrated for the smart healthcare device. However, as situations become complex, large number of propositional variables there would be a need to sufficiently express EOP. This would increase the possible number of scenarios exponentially and it may be difficult to work out the manners map byhand. Though focusing on context specific ethical requirements may fend off the problem to some extent. A mechanism that is able to automate the generation of the truth tables from a given EOP would be helpful in such situations.

Although smart things are expected be endowed with some processing power, analyzing large exponential scenarios may not be feasible on such devices. Under such circumstances, fog and cloud computing platform solutions may need to be explored. (Fog computing provides networking, storage and computing facilities to IoT nodes locally within the network in order to reduce latency associated with cloud computing.) As the numbers of IoT devices continue to increase exponentially cloud computing has the potential to address several IoT requirements such as cost, scalability and data analytics. The related concept of fog computing brings cloud closer to the devices in the network, as well as addressing the issues of privacy latency and bandwidth. A detailed discussion of fog and cloud computing with regard to IoT is given by Dizdarević et al. (2019). Also, once ethical parameters have been sufficiently represented in smart devices they would also need to be secured. The question of ethical dilemmas in moral philosophy also represents a considerable challenge. However, it may not be wise to halt engineering efforts toward designing ethics in the face of dilemmas that have eluded philosophers for centuries. For such problems, 'human in the loop' mechanisms can be explored. Although we have sketched a possible solution for incorporating ethics in smart devices, considerable research effort would be needed to realize a flexible approach toward ethics implementation.

9.6 Conclusion

The novel paradigm of IoT has the potential to transform modern societies. Due to wide scope of smart application services, IoT has promising technological, economic and social prospects. However, with increased pervasiveness and autonomy of smart devices, ethical conformity becomes essential. In spite of the large scale research effort, ethical considerations of the technology have not been adequately addressed.

In this work, we have proposed a novel methodthat is able to implement context specific ethics or EOP in smart devices. EOP expressed in terms of propositional variables are mapped to ethically consistent values in the manners map. The method was illustrated by considering a smart healthcare device that informs family and an authorized doctor about the health condition of a patient in an ethically acceptable manner. By extending a similar approach to other smart devices, we can ensure that smart devices function in a manner congruent to ethical expectations of society. We hope this work would further the research effort toward incorporating ethics in smart devices.

References

Allen, Colin, Iva Smit, and Wendell Wallach. 2005. "Artificial Morality: Top-down, Bottom-up, and Hybrid Approaches." *Ethics and Information Technology* 7 (3): 149–155. doi:10.1007/s10676-006-0004-4.

Anderson, Michael, and Susan Leigh Anderson. 2011. *Machine Ethics*. Cambridge University Press, Cambridge, United Kingdom.

Anderson, Michael, Susan Leigh Anderson, and Chris Armen. 2005. "Towards Machine Ethics." *Proceedings of the AAAI 2005 Fall Symposium on Machine Ethics*, Menlo Park, California.

Anderson, Michael, Susan Leigh Anderson, and Chris Armen. 2006. "MedEthEx : A Prototype Medical Ethics Advisor." In *Proceedings of the 18th Conference on Innovative Applications of Artificial Intelligence*, Volume 2, 1759–1765.

Baldini, Gianmarco, Maarten Botterman, Ricardo Neisse, and Mariachiara Tallacchini. 2016. "Ethical Design in the Internet of Things." *Science and Engineering Ethics*, 24: 905–925. doi:10.1007/s11948-016-9754-5.

DizdarevićJasenka, FranciscoCarpio, Admela Jukan, and Xavi Masip-Bruin. 2019. "A Survey of Communication Protocols for Internet of Things and Related Challenges of Fog and Cloud Computing Integration." *ACM Computing Surveys* 51 (6): 1–29.

Friedman, B., Kahn Jr., P. H., & Borning, A. (2006). Value Sensitive Design and Information Systems. In *Human-Computer Interaction and Management Information Systems: Foundations* (pp. 348–372). M. E. Sharpe, Inc., New York, United States

Jasanoff, S. (2016). *The Ethics of Invention: Technology and the Human Future*. WW Norton & Company, New York, United States.

Rosen, Kenneth H. 2009. *Discrete Mathematics and Its Applications*. McGraw-Hill, New York, United States.

Sholla, Sahil, Roohie Naaz Mir, and Mohammad Ahsan Chishti. 2018. "Eventuality of an Apartheid State of Things: An Ethical Perspective on the Internet of Things." *International Journal of Technoethics* 9 (2): 62–76.

Wallach, Wendell, and Colin Allen. 2009. *Moral Machines: Teaching Robots Right from Wrong*. Oxford University Press, Oxford, United Kingdom.

10

Security Solutions for Threats in IoT-Based Smart Vehicles

Surbhi Sharma and Baijnath Kaushik

CONTENTS

10.1 Introduction

IoT is assumed as an innovative transformation as it connects the physical and digital world (Ray, 2018). IoT enables the communication between different things (humans, objects, etc.) through the Internet, and that is why it is named 'the internet of things' resulting in a smarter and more intelligent planet. The concept of IoT has led to the vision of "anytime, anywhere, anyway and anything communications" as everyday devices are becoming smarter, everyday processing is becoming intelligent, and everyday communication is becoming informative (Feki et al., 2013). IoT has its applications in numerous areas, and thus it has open the research opportunities in various sectors by incorporating the concept of smartness in multiple domains like smart-home, smart-industry, smart-transport and smart-health (Atzori et al., 2010; Kaiwartya et al., 2016; Sharma and Kaushik, 2019) as represented in Figure 10.1.

Deployment of IoT in transportation systems leads to the emergence of intelligent transportation systems (ITS) resulting in the concept of IoV which have the composition of VANETS (vehicular ad hoc networks) and IoT. Due to the incorporation of IoT in transportation systems, IoT-based vehicles (IoV) can efficiently handle traffic-related issues as well as enhances the driving experience thus resulting in the safety of passengers, which is the main objective of ITS (Nundloll et al., 2009; Sharma and Kaul, 2018a). The term 'IoT-based smart vehicles' and 'IoV' can be used interchangeably. Figure 10.2 shows the composition of IoV.

As everything comes with pros and cons, thus, despite numerous benefits of IoT, its security aspects need to be focused because security cannot be compromised in any scenario. The main objective of IoT is 24–7 communication between different objects due to it, a vast amount of safety-critical data is generated but IoT devices are resource-constrained

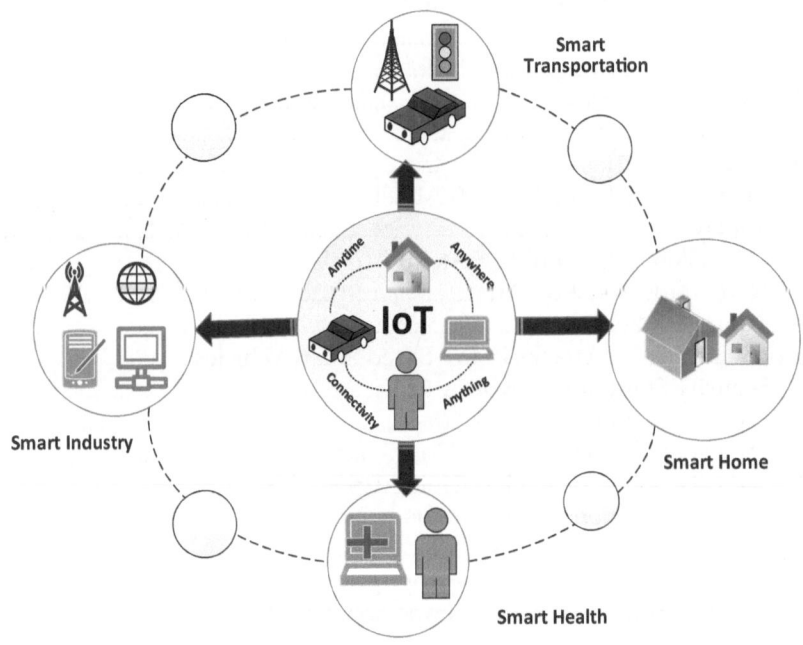

FIGURE 10.1
Application of IoT in different sectors.

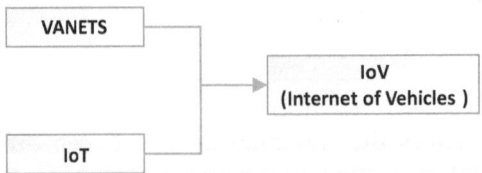

FIGURE 10.2
Composition of IoV.

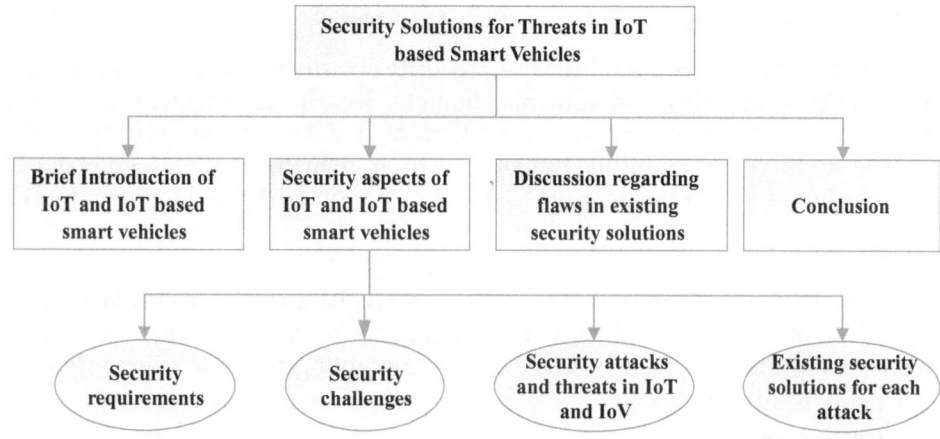

FIGURE 10.3
Structure of chapter.

devices (i.e., limited storage, computing and network capacities) thus, IoT devices are exposed to various kinds of cyber-attacks. In IoT-based smart vehicles security also needs to be focused, otherwise the hacker can hack vehicles because everything is connected to the Internet, resulting in hazardous situations that could lead to the loss of lives (Ali et al., 2019). Thus, it is of utmost importance to do an extensive survey on different security attacks possible in IoT and the IoV and existing security solutions.

Main contributions of this chapter

The main contributions of this chapter are:

1. A detailed survey on IoT and IoT-based smart vehicles (IoV), taking into consideration its security aspects.
2. Security is a major concern in IoT, so security requirements is studied, focusing on security goals like confidentiality, authenticity, etc.
3. Security challenges are also focused on in this survey as they need to be identified before the actual deployment of any framework.
4. An in-depth study on security attacks in IoT and IoV are done, which are mainly categorized into active and passive attacks.
5. This study also focuses on security solutions proposed for each attack.
6. Highlighting the shortcomings of existing proposed security solutions, and focus on other attacks that haven't gained much attention, for which solutions are not proposed yet.

Figure 10.3 illustrates the workflow of the chapter (i.e., topics to be addressed).

10.2 Security Aspects of IoT and IoT-Based Smart Vehicles

Security is of utmost importance in IoV as it can lead to misleading actions when a vehicle is controlled by a hacker. Thus, it is important to take into consideration security aspects in depth, including security requirements, security challenges, security attacks and are elaborated in the following section.

10.2.1 Security Requirements in IoT-Based Smart Vehicles

Security requirements are acquired from basic security objectives like confidentiality, access control, integrity, availability and authentication etc. Security requirements are of prime concern before deploying any security framework as it elaborates the security level of a network and these should be fulfilled in all security frameworks (Al-Qutayri et al., 2010; Zhang and Delgrossi, 2012). Security requirements are explained in the following section:

i) *Confidentiality:*

It ensures that information should remain secret while transmitting between vehicles as IoV involves real-time data, thus, information is critical in such vehicular networks. Confidentiality can be achieved by encrypting the data (Bernardini et al., 2017; Zaharuddin et al., 2010)

ii) *Authentication:*

It aims to ensure that only legitimate vehicles should be involved in a communication otherwise legitimate nodes might be controlled by malicious vehicle for broadcasting the messages in the network for its benefit resulting in hazardous actions (Daeinabi and Rahbar 2013; Guo et al., 2007).

iii) *Data Integrity:*

IoV involves safety-critical data because any change in content can result in loss of lives. Data integrity ensures accuracy and consistency of information, and it prevents malicious users from accessing any information like the vehicle's identity, vehicle's location etc. To maintain data integrity, digital signatures along with password can be used (Mokhtar and Azab, 2015).

iv) *Availability:*

It ensures that legitimate users should have timely information whenever it is required otherwise the information is of no worth as IoV involves real-time-sensitive information. To maintain the availability of information, Group signature scheme has been used widely (Guo et al., 2007).

v) *Non-repudiation:*

Its goal is to avoid the denial of sender or receiver regarding broadcasted information (Tangade and Manvi, 2013).

vi) *Access control:*

In IoV network, different roles and privileges are assigned to all participating vehicles so that effective communication can take place. Access control ensures that participating vehicles should perform tasks which they are supposed to do (Sumra et al., 2011a).

10.2.2 Security Challenges in IoT-Based Smart Vehicles

Before the actual deployment, it is required to identify the security challenges of IoT and IoV. Few of the challenges are- High mobility, Low error tolerance, Key management, a tradeoff between privacy and security etc. and are discussed below:

i) *High mobility:*

As IoV network is highly mobile due to all-time movement of vehicles but these dynamic topological structures hamper the V2V and V2R communications due to it, security and non-repudiation can be compromised (Qu et al., 2015).

ii) *Low error tolerance:*

IoV network involves real-time data, any delay or other minor mistake can lead to hazardous results. Thus, to prevent accidents, drivers need to be warned in advance as a precautionary measure (Aouzellag et al., 2015).

iii) *Key management:*

Cryptography includes keys for encrypting and decrypting the sensitive data thus key management is one of the important challenges which need to be focused while designing the security protocols (Golle et al., 2004).

iv) *Tradeoff between privacy and security:*

Security is of utmost importance in a vehicular network like IoV and VANETs as security can't be economized on in these ITS, otherwise the vehicle may be hacked by a malicious vehicle resulting in misleading actions. More secure system has chances of less privacy but, at the same time, privacy also need to be taken care of. Thus there need to be a tradeoff between good performance and security and is a major challenge (Malik and Bishnoi, 2014; Wu et al., 2016).

10.3 Security Attacks and Threats in IoT and IoT-Based Smart Vehicles

IoT-based vehicles are vulnerable to numerous security attacks. Existing security attacks in IoV need to be studied to enhance security. This section will focus on possible active and passive attacks in IoT and IoV. In active attack, malicious user intercepts the network, modifies the data or broadcasts false messages. Message replay is an example of an active attack where malicious users deliberately forward the same message multiple times to perform misleading action. In contrast, in the passive attack, attacker passively analyzes the sensitive information from the network rather than directly modifying it. Traffic analysis is an example of a passive attack in which an attacker silently analyzes the traffic pattern. Bifurcation of attacks to be discussed are mentioned below in Table 10.1.

TABLE 10.1

Categories of Security Attacks in IoT and IoV

Active Attacks		Passive Attacks
Sybil Attack	Spamming Attack	Repudiation Attack
Sink Hole	Message Tampering Attack	ID Disclosure
Session Hijacking	GPS Spoofing	Session Hijacking
Home Attack	Man-in-the-Middle attack	Wormhole
Distributed Denial of Service	Black Hole	Snooping
Illusion Attack	Replay Attack	
Node Impersonation Attack	Masquerading	
Wormhole	Social Attack	
Denial of Service	Gray Hole	
Brute Force Attack	Timing Attack	
Key and/or Certificate		

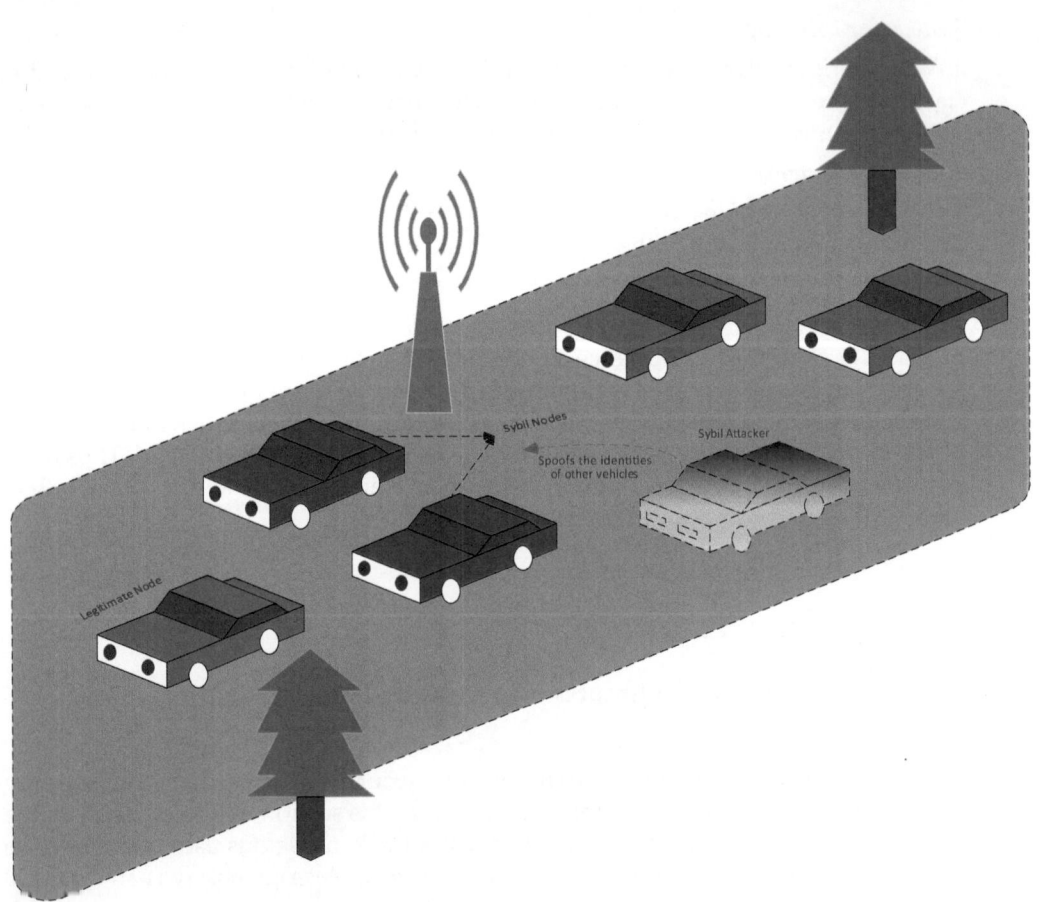

FIGURE 10.4
Sybil attack.

10.3.1 Sybil Attack

In Sybil attack, multiple identities of malicious node exist thus, it is critical to distinguish between the malicious node and legitimate node. Huge security risk is involved in a network due to the presence of multiple identities as vehicles may assume information is from reliable vehicle which is being sent by malicious vehicle resulting in loss of lives. Vehicle spoofing identities of others are known as a Sybil attacker and vehicles whose identities are spoofed are known as Sybil nodes (Guette and Ducourthial 2007; Sumra et al., 2011b). Figure 10.4 Illustrates Sybil attack in a network.

Sybil attack lowers the performance of the network by consuming the resources (Bariah et al., 2015). It can be performed in three ways- Based on communication type, type of participation in a network and Sybil identity type (Sharma and Kaul 2018b).

10.3.2 Denial of Service (DoS) Attack

DoS attack violates the availability security goal. It disrupts the communication between vehicles by preventing legitimate vehicles from accessing the network resources (Rawat

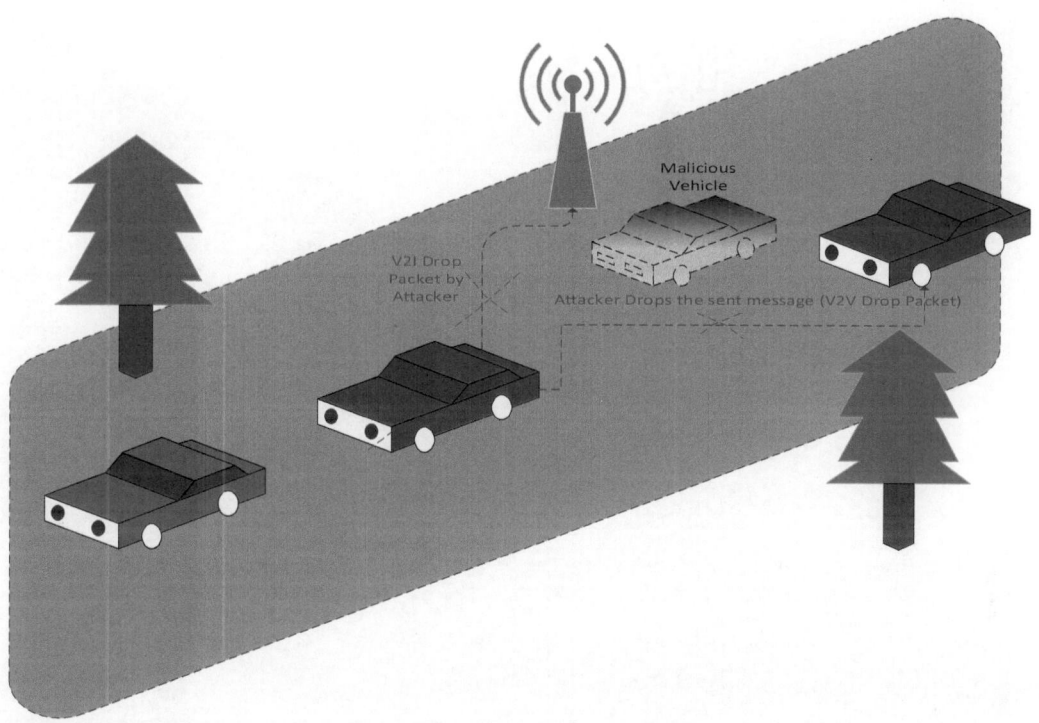

FIGURE 10.5
DoS attack between V2I and V2V due to Packet drop.

et al., 2012). In safety-critical applications, any kind of delay is intolerable as communication is most important (Malla and Sahu, 2013).

DoS attack has three main properties: remote, malicious and disruptive (Kim et al., 2014). Three different ways for performing DoS attack are: packet dropping, network overloading and communication channel jamming (Hasbullah and Soomro, 2010; Sumra et al., 2011a). Figure 10.5 shows the example of a DoS attack between V2V and V2I.

10.3.3 Black Hole Attack

This attack relies on the concept of black hole. Entire traffic of network is routed to an area known as a black hole either due to packet drop in between or due to denial of participation. In this attack, a malicious node introduces itself as a node with the shortest path to destination thus entire traffic is routed through black hole area where data is altered before forwarding it to the destination node (Abdulkader et al., 2017; Dhamgaye and Chavhan, 2013).

Figure 10.6 shows an example where malicious vehicles are performing black hole attack.

10.3.4 Gray Hole Attack

It is a variant of black hole attack and in this attack, malicious vehicles have dual nature (i.e., in certain interval it behaves as an innocent node and, in another interval, it acts as malicious node). This attack is difficult to detect and it aims to lower the packet delivery

FIGURE 10.6
Black hole attack.

ratio of network as malicious nodes selectively drops the drop packets and rest are forwarded unlike black hole attack (Hasrouny et al., 2017; Nogueira et al., 2012; Verma et al.,
2015).

Figure 10.7 illustrates example of gray hole attack between V2I and V2V.

10.3.5 Wormhole Attack

In this attack, malicious nodes build a tunnel known as wormhole. For creation of a tunnel,
at least two malicious nodes participate where malicious node at one end of the tunnel
routes the entire traffic to other malicious nodes available at another end of tunnel. Security
of data packets is thus compromised (Abdulkader et al., 2017; Bendjima and Feham, 2016)

Figure 10.8 explains the example of wormhole attack.

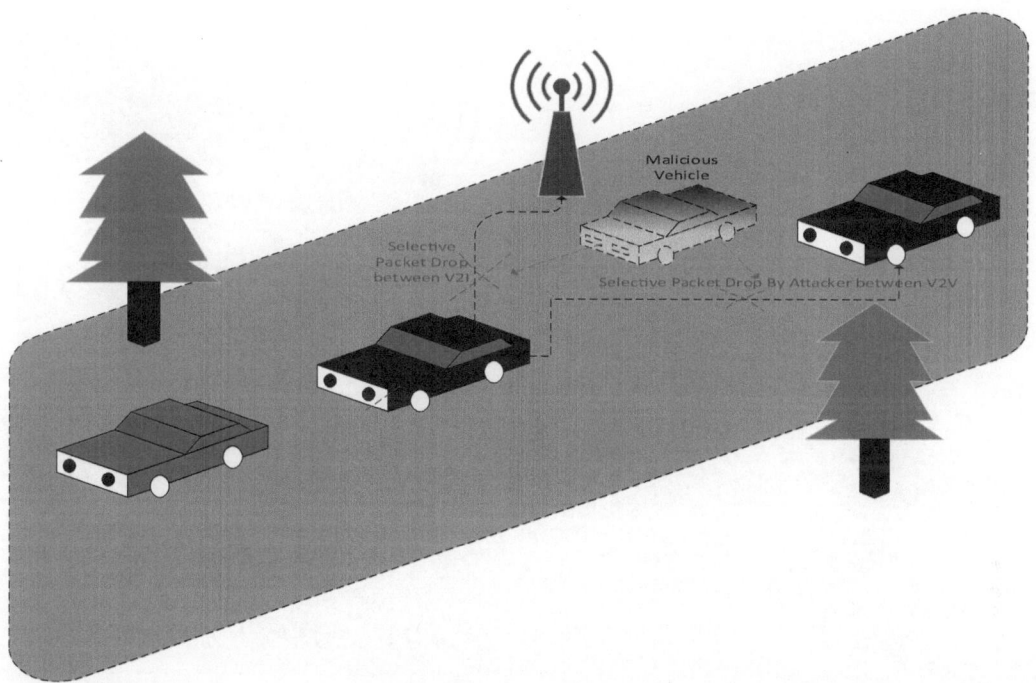

FIGURE 10.7
Gray hole attack.

10.3.6 Man-in-the-Middle (MITM) Attack

Main aim of this attack is to breach privacy and data integrity. In this attack, a malicious vehicle is present in between two communicating vehicles where it eavesdrops the entire communication for its own benefits while communicating legitimate vehicles have no clue about it. It has a negative impact on the authenticity of transmitted information (Kumar and Sinha, 2014; Lipiński et al., 2015; Malla and Sahu, 2013). Figure 10.9. explains the MITM attack where information is modified by malicious vehicle.

10.3.7 Replay Attack

Replay attack is an active attack in which a malicious node or attacker broadcasts the same message repeatedly, which has already been forwarded to the destination vehicle. The main aim is to increase the bandwidth cost due to frequent replaying of messages and to drop the priority messages (Sakiz and Sen, 2017; Zeadally et al., 2012). Figure 10.10 Illustrates an example of a replay attack where malicious vehicle sends the information with some delay to neighboring vehicles.

10.3.8 GPS (Global Positioning System) Attack

Having GPS system in vehicles helps with determining the vehicle's location, and is based on satellites. In a GPS spoofing attack, the attacker uses stronger signals with the help of GPS simulators, and thus manipulates the vehicle's position (Abdulkader et al., 2017). In this process, fake locations will be forwarded to other vehicles, which in turn will disrupt the entire communication, potentially resulting in accidents (Hasrouny et al., 2017).

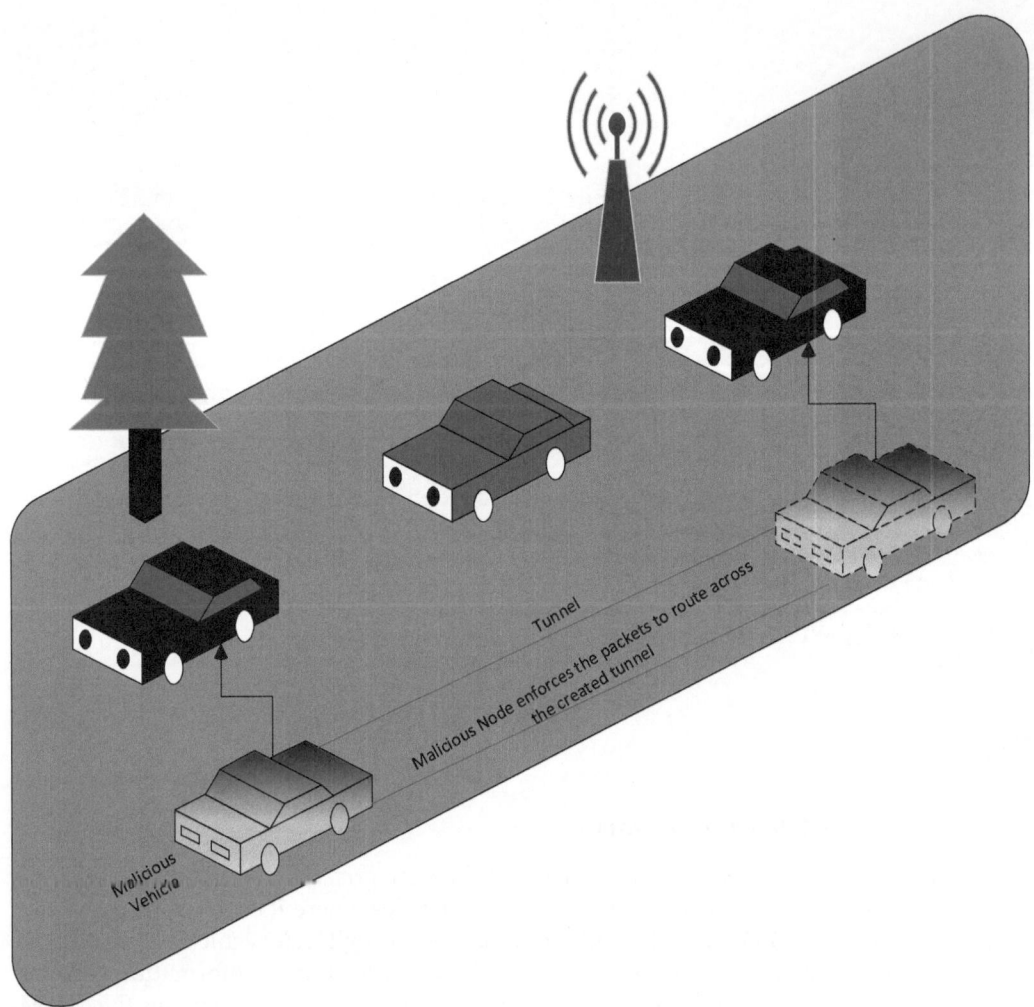

FIGURE 10.8
Wormhole attack between vehicles.

10.3.9 Node Impersonation Attack

Node impersonation attack poses a threat to authentication. In this attack, a malicious vehicle changes its unique ID for performing a misleading task. Data is modified after changing the ID and modified data is broadcast to neighboring vehicles, and confidentiality is violated as neighboring vehicles then assume that received information is from legitimate sender node (Mejri et al., 2014).

10.3.10 Timing Attack

As IoV have numerous safety-critical applications where time plays an important role. Timely delivery of information to other vehicles is of utmost importance, because any delay makes the information useless. In a timing attack, a malicious user adds a time slot to the original message rather than deliberately modifying the message resulting in late delivery of message which is worthless (Sumra et al., 2011a).

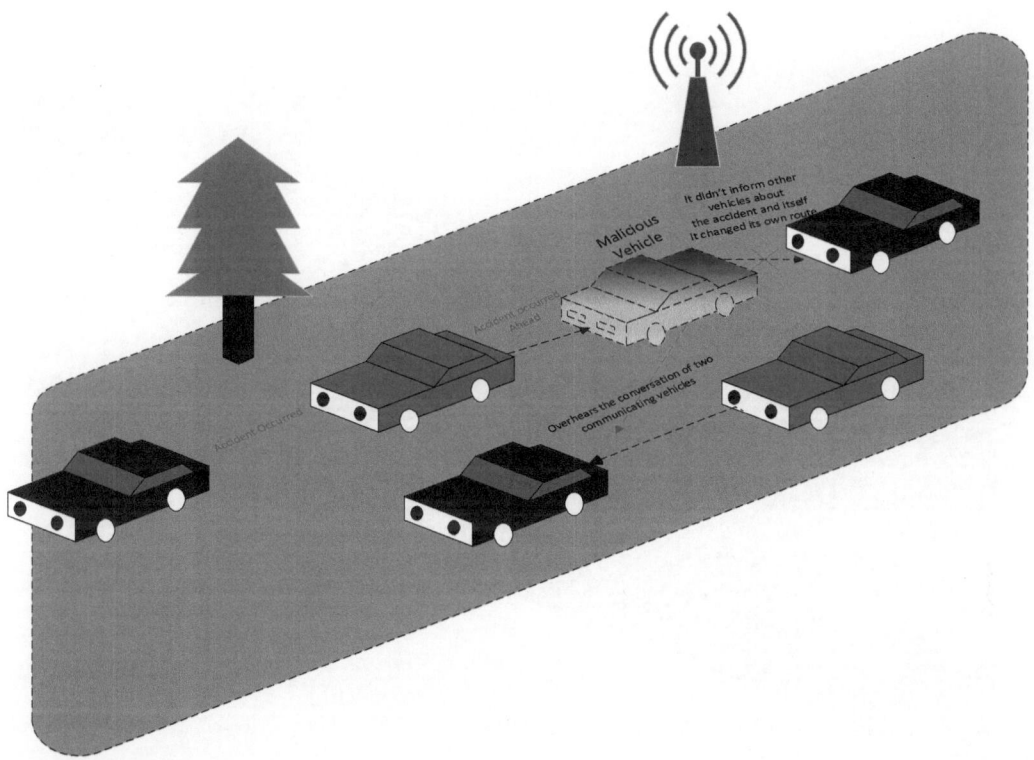

FIGURE 10.9
MITM attack in IoV.

10.4 Security Solutions for Attacks in IoT-Based Smart Vehicles

This section focuses on an exhaustive survey on the existing security solutions proposed for each attack, on the advantages and disadvantages, as well as simulators used in each proposed solution.

10.4.1 Security Solutions to Sybil Attack

The Sybil attack is one of the important attacks in vehicular networks like IoV, and it affects the entire communication between vehicles as multiple identities of each node exists. Thus, it is important to be aware of security solutions for Sybil attacks. A Sybil attack detection method with obfuscated RSUs vector-based signature scheme has been proposed by Feng and Tang (2017). The proposed method relies on a few considerations: ring signature-based identification scheme has been designed to identify malicious Sybil nodes; preserving the vehicle's private information by baffling the RSUs neighboring relationship; achieving low overhead as well as high efficiency; obtaining online and independent detection. **Advantages**: lower computational overhead and higher detection rate than other existing methods. **Disadvantages:** in signature verification, information synchronization of RSUs exists. **Future Scope**: considering local information, identity authentication will be conducted. **Simulators used**: SUMO and MOVE. **Evaluation Parameters**: detection rate and computational overhead.

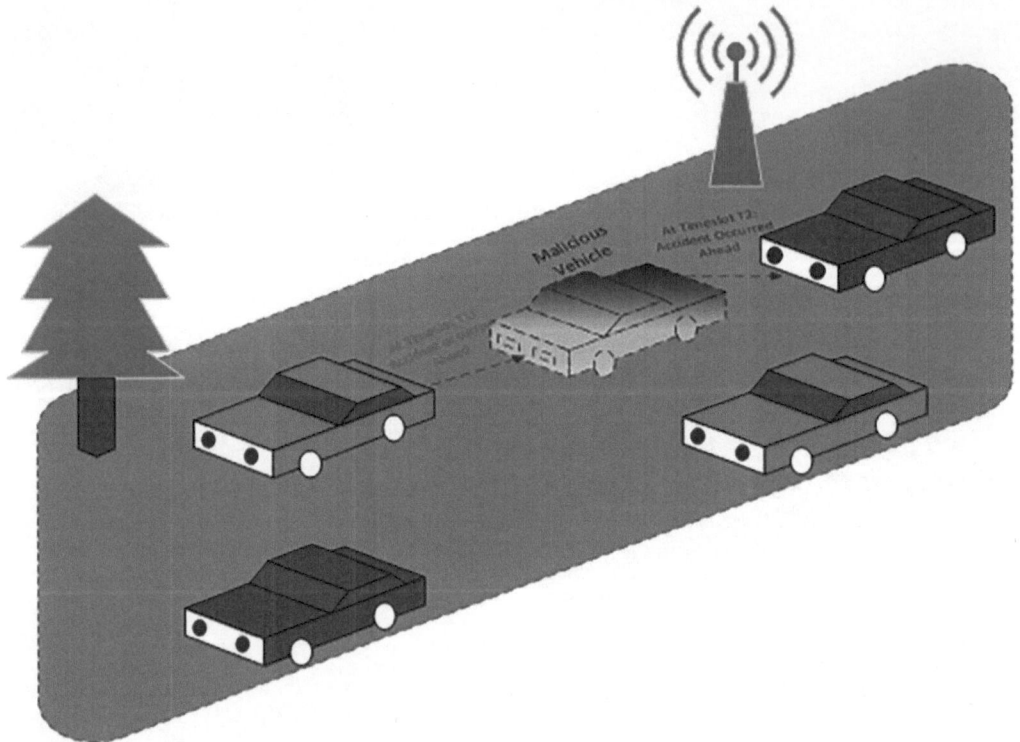

FIGURE 10.10
Replay attack.

Feng et al., (2017) have designed a method against multi-source Sybil attacks known as Event-based Reputation System (EBRS). Local certificate generation and local certificate validation are the two main steps in the proposed method. Time-limited pseudonyms are taken into consideration to maintain the privacy of vehicles as well, and it relies on each event's unique reputation and trusted value to detect multi-source malicious Sybil nodes. **Advantages**: EBRS method ensures privacy of nodes and prevents multi-source Sybil attacks. **Future Scope**: RSU's strong assumption will be detached. **Simulators used**: NS-2. **Evaluation Parameters**: delivery ratio and communication delay.

10.4.2 Security Solutions to DoS Attack

The main objective of DoS attack is to prevent legitimate users from accessing the services to which they are entitled to. Enhanced Attacked Packet Detection Algorithm (EAPDA) has been designed by Singh and Sharma (2015) as a solution to prevent DoS attack. In this method, RSU plays a key role as each RSU verifies the rate of sending data packets from each node/vehicle. The malicious vehicle will be identified once RSU notices strange behavior of sending packets and that node will be eliminated. **Advantages**: detects the malicious node in the verification stage so, it prevents the delay as well as, false-positive rate is boosted up. **Future Scope**: assigning priority and verifying emergency vehicles quickly. **Simulator used**: NS-2. **Evaluation Parameters**: delay, false-positive rate and throughput.

Two-phase signature-based authentication scheme has been proposed by Pooja et al. (2014) to prevent insider and outside DoS attacks. The main aim of Phase 1 is to alleviate outsider DoS attacks and is based on Hash-based message authentication codes. Phase 2 is to detect inside attackers and it is performed after the successful completion of Phase 1. In Phase 2, each vehicle builds its two tables-blacklist table and detection table to detect insider attackers. **Advantages**: computational overhead is negligible. **Disadvantages**: proposed system can't detect bogus information. **Simulator used**: NS-2. **Evaluation Parameters**: computational overhead.

10.4.3 Security Solutions to Black Hole Attack

Black hole attack poses a threat to availability. Cherkaoui et al. (2017) have designed a quality control chart method to avoid black hole attack. Statistical Process Control (SPC) is taken as baseline in the proposed method. To represent the abnormal behavior of malicious nodes/vehicles, quality control charts are used named as p-charts. P-charts are used in analyzing the network's lost traffic to identify malicious node causing black hole attack. **Advantages**: no additional development is required, thus it is better and efficient as well as abnormal behavior are detected in real-time. **Simulator used**: NS-2. **Evaluation Parameters**: packet loss ratio.

Authors in (Tobin et al., 2017) have designed an approach for detecting black hole attacks assuming the presence of a single malicious node and others are loyal nodes. Attack detection, node accusation and malicious node blacklisting are the three main stages in the proposed approach. Attack detection relies on route backtracking and, its aim is to notice the discrepancies announced by the neighboring vehicles. In node accusation stage, source and destination nodes accuse the node as a malicious one once being detected in the previous stage and forward this information to neighbors as well. In the last stage, nodes which receive accusation messages build blacklist entries, and all blacklist entries are kept in a 'black chain' and the malicious node won't then be able to affect the entire communication **Advantages**: effective in detecting and eliminating a single malicious node from the network. **Disadvantages**: only single malicious node is assumed to be present in a proposed method which is not possible in real scenarios. **Simulator used**: NS-3. **Evaluation Parameters**: accuracy of detection and detection time, packet loss, average throughput and packet delivery ratio (PDR).

10.4.4 Security Solutions to Gray Hole Attack

Gray hole is difficult to detect due to the dual nature of nodes. Ali Alheeti et al. (2016) have designed an intelligent intrusion detection system to prevent gray hole attacks. Feed Forward Neural Networks (FFNN) and Support Vector Machines (SVM) are the baseline methods for the proposed approach. Proportional Overlapping Score (POS) is used as a criterion to extract desired features and lowest weight features are eliminated. **Advantages**: proposed method reduces false alarms and have high detection rates. **Simulator used**: NS-2. **Evaluation of Parameters**: false alarm rate and detection accuracy.

Kaur (2016) designed a genetic-based algorithm to alleviate gray hole attack. Fitness function is used to identify gray hole nodes and then optimization is performed using Artificial Bee Colony optimization (ABC) and genetic algorithm. **Advantages**: results are good because genetic algorithms take into consideration global optima rather than local optima. **Simulator used**: MATLAB. **Evaluation Parameters**: throughput, Bit Error Rate, PDR and End-to-End Delay.

10.4.5 Security Solutions to Wormhole Attack

Wormhole attacks are difficult to detect due to the presence of private tunnel, which is invisible to the network. Ali et al. (2017) have proposed a cryptographic based technique against wormhole attacks. RSA algorithm—which is an asymmetric cryptographic algorithm used for distribution of the shared key for transmitting the message in the network symmetric key cryptographic algorithm—is used as messages need to be securely transmitted, and the message contains important information (i.e., time at which data is sent, id of the sender node and location of sender). **Advantages**: energy of nodes is saved due to the consideration of both shared key and public-key encryption. Along with this, the proposed method includes fewer computations thus, message broadcasting is secure. **Disadvantages**: high consumption of power as the number of nodes increases.

Heuristic-based approach is suggested by Nikam and Sarawagi (2017) to mitigate wormhole attacks. In this approach, elliptic curve cryptography (ECC) is used over AODV routing protocol. Along with this, heuristics is used to decide the path of data packet and node with high-speed is nominated as energy head. **Advantages**: efficient in preventing wormhole attacks and toward packet transfer. **Simulator used**: NS-2. **Evaluation Parameters**: throughput, PDR and average delay.

10.4.6 Security Solutions to MITM Attack

In this attack, a malicious node acts as an intermediate between two communicating vehicles. Anonymous Location-Based Efficient Routing Protocol (ALERT) has been modified by Patil et al. (2016) to prevent MITM attack, and it relies on location-based pseudonyms and hash-functions (e.g., SHA-1). It relies on ACK (acknowledgment to check whether the data packets are sent successfully or not). Hash values are calculated at source and destination nodes and negative acknowledgment is forwarded when malicious nodes are detected, and thus alternate route is chosen for message forwarding. **Advantages**: efficient and feasible approach for reliable communication. **Disadvantages**: time-consuming approach due to simulation used for sending data. **Simulator used**: JAVA using JUNG tools.

10.4.7 Security Solutions to GPS Spoofing Attack

In a GPS spoofing attack, the safety of passengers is compromised due to fake GPS signals leading them to wrong locations. Two-factor authentication method is suggested by Tayeb et al. (2017) to prevent GPS spoofing attack. A digital signature (i.e., RSA-1024) is used to hash the GPS signal, then a hashed signal is encrypted using a private key, and ultimately, the public key is used to decrypt the data. **Advantages**: safety of passengers is guaranteed and it is a feasible approach. **Future scope**: preventing the hijacking of cars by proposing a more secured approach. **Materials used**: computer as a sender, a Stratum-1 time server along with NTP version 4 and Raspberry Pi as a receiver.

10.4.8 Security Solutions to Node Impersonation Attack

Malicious attacker achieves unauthorized access to the network privileges in node impersonation attack and, thus the network is disturbed. A framework has been designed by Gour and Kumar (2014) to prevent node impersonation attack. It takes into consideration the spatial association of RSS for assault identification. It is independent of cryptography, and RSS hub character is split into two classes by using unsupervised threshold methodology. **Advantages**: adversaries from the network are eliminated effectively. **Simulator used**: NS-2. **Evaluation Parameters**: throughput, End-to-End Delay and PDR.

10.5 Discussions

Based on our exhaustive survey on security threats and security solutions, it is analyzed that existing security solutions have numerous flaws which need to be addressed further. Table 10.2 summarizes the attacks for which security solutions are not yet proposed and outlines the flaws of existing security solutions deployed for each attack.

TABLE 10.2

Drawbacks of Existing Proposed Security Solutions

Attack Name	Security Goal Violated	Drawbacks in Existing Proposed Security Solutions	Remarks
Sybil attack	Confidentiality and Availability	Dependency on RSUs, Dependency on the number of nodes etc.	Scope for further enhancement
DoS Attack	Availability	Depends on manual settings of parameters thus, not a feasible option, Inability to detect malicious node in case network have bogus information and valid signatures	Scope for further improvement due to mentioned flaws
Black Hole Attack	Availability	Few are computationally heavy and needs extra resources, others take into consideration the presence of only a single malicious node which is not possible practically	Need for further enhancement to overcome the existing flaws
Gray Hole Attack	Availability	Only numerical values as input are acceptable by few proposed methods, few are less efficient due to dependency on traditional routing protocols	Need to consider cryptographic methods to secure the network from such attacks
Wormhole Attack	Confidentiality & Availability	High consumption of power due to the number of nodes as well as comparative results are not found	Scope for further enhancement to secure the network
Node Impersonation Attack	Authentication	High overhead and delay, Not much secure	Not much security solutions have been yet proposed for these attacks
MITM attack	Integrity, Confidentiality and Availability	Time-consuming approach due to simulation used for sending data	Not much security solutions have been yet proposed for these attacks
GPS Spoofing Attack	Integrity and Availability	Less secure	Not much security solutions have been yet proposed for these attacks
Replay Attack and Timing Attack	Integrity and Availability	Not much work was found on these attacks in the literature	Not much security solutions have been yet proposed for these attacks; thus, more security solutions need to design for these attacks to secure the network

10.6 Conclusion

IoT is a domain that represents the next most exciting technological revolution since the Internet. In IoT, key technologies which are primarily used are RFID, sensor networks, real-time localization, and short-range wireless communications. IoT devices are resource-constrained devices and are easy to hack and compromise. So, the security of IoT and IoT-based smart vehicles are being focused in this chapter. Security is a prime concern in IoT-based vehicles also, as vehicles can behave abnormally in case, it is hacked by hackers leading to loss of lives thus it cannot be ignored and is of prime concern. The extensive survey is being carried out in this proposal taking into consideration all security aspects (i.e., security requirements: all security goals like confidentiality, authenticity, etc.), security challenges, possible security attacks (active and passive attacks). After this, proposed security solutions for each attack are also focused as well as existing security solutions have a lot of shortcomings that need to be focused, and security solutions need to be proposed for other attacks.

References

Abdulkader, Zaid A, Azizol Abdullah, Mohd Taufik Abdullah, and Zuriati Ahmad Zukarnain. 2017. "Vehicular ad hoc networks and security issues: survey." *Modern Applied Science* 11 (5):30.

Al-Qutayri, Mahmoud, Chan Yeun, and Faisal Al-Hawi. 2010. "Security and privacy of intelligent VANETs." In *Computational Intelligence and Modern Heuristics, Al-Dahoud Ali (Ed.)*. IntechOpen, Jordan.

Ali Alheeti, Khattab M, Anna Gruebler, and Klaus McDonald-Maier. 2016. "Intelligent intrusion detection of grey hole and rushing attacks in self-driving vehicular networks." *Computers* 5 (3):16.

Ali, Ikram, Alzubair Hassan, and Fagen Li. 2019. "Authentication and privacy schemes for vehicular ad hoc networks (VANETs): A survey." *Vehicular Communications* 16, 45–61.

Ali, Shahjahan, Parma Nand, and Shailesh Tiwari. 2017. "Secure message broadcasting in VANET over Wormhole attack by using cryptographic technique." *2017 International Conference on Computing, Communication and Automation (ICCCA)*, Greater Noida, India.

Aouzellag, Ha roune, Kaci Ghedamsi, and Djamel Aouzellag. 2015. "Energy management and fault tolerant control strategies for fuel cell/ultra-capacitor hybrid electric vehicles to enhance autonomy, efficiency and life time of the fuel cell system." *International Journal of Hydrogen Energy* 40 (22):7204–7213.

Atzori, Luigi, Antonio Iera, and Giacomo Morabito. 2010. "The internet of things: A survey." *Computer Networks* 54 (15):2787–2805.

Bariah, Lina, Dina Shehada, Ehab Salahat, and Chan Yeob Yeun. 2015. "Recent advances in VANET security: a survey." *2015 IEEE 82nd Vehicular Technology Conference (VTC2015-Fall)*, Boston, MA, USA.

Bendjima, Mostefa, and Mohammed Feham. 2016. "Wormhole attack detection in wireless sensor networks." *2016 SAI Computing Conference (SAI)*, London, UK.

Bernardini, Cesar, Muhammad Rizwan Asghar, and Bruno Crispo. 2017. "Security and privacy in vehicular communications: Challenges and opportunities." *Vehicular Communications* 10:13–28.

Cherkaoui, Badreddine, Abderrahim Beni-Hssane, and Mohammed Erritali. 2017. "Quality control chart for detecting the black hole attack in vehicular Ad-Hoc networks." *Procedia Computer Science* 113:170–177.

Daeinabi, Ameneh, and Akbar Ghaffarpour Rahbar. 2013. "Detection of malicious vehicles (DMV) through monitoring in Vehicular Ad-Hoc Networks." *Multimedia Tools and Applications* 66 (2):325–338.

Dhamgaye, Anup, and Nekita Chavhan. 2013. "Survey on security challenges in VANET 1."

Feki, Mohamed Ali, Fahim Kawsar, Mathieu Boussard, and Lieven Trappeniers. 2013. "The internet of things: the next technological revolution." *Computer* (2):24–25.

Feng, Xia, Chun-yan Li, De-xin Chen, and Jin Tang. 2017. "A method for defensing against multi-source Sybil attacks in VANET." *Peer-to-Peer Networking and Applications* 10 (2):305–314.

Feng, Xia, and Jin Tang. 2017. "Obfuscated RSUs vector based signature scheme for detecting conspiracy Sybil attack in VANETs." *Mobile Information Systems 2017*.

Golle, Philippe, Dan Greene, and Jessica Staddon. 2004. "Detecting and correcting malicious data in VANETs." *Proceedings of the 1st ACM international workshop on Vehicular ad hoc Networks*, Philadelphia, PA, USA.

Gour, Nidhi, and Ajay Kumar. 2014. "Efficient Detection and Prevention of Impersonation attack in MANET."

Guette, Gilles, and Bertrand Ducourthial. 2007. "On the Sybil attack detection in VANET." *2007 IEEE International Conference on Mobile Adhoc and Sensor Systems*, Pisa, Italy.

Guo, Jinhua, John P Baugh, and Shengquan Wang. 2007. "A group signature based secure and privacy-preserving vehicular communication framework." *2007 Mobile Networking for Vehicular Environments*, Anchorage, AK, USA.

Hasbullah, Halabi, and Irshad Ahmed Soomro. 2010. "Denial of service (dos) attack and its possible solutions in VANET." *International Journal of Electronics and Communication Engineering* 4 (5):813–817.

Hasrouny, Hamssa, Abed Ellatif Samhat, Carole Bassil, and Anis Laouiti. 2017. "VANet security challenges and solutions: A survey." *Vehicular Communications* 7:7–20.

Kaiwartya, Omprakash, Abdul Hanan Abdullah, Yue Cao, Ayman Altameem, Mukesh Prasad, Chin-Teng Lin, and Xiulei Liu. 2016. "Internet of vehicles: Motivation, layered architecture, network model, challenges, and future aspects." *IEEE Access* 4:5356–5373.

Kaur, Gurleen. 2016. "A preventive approach to mitigate the effects of gray hole attack using genetic algorithm." *2016 International Conference on Advances in Computing, Communication, & Automation (ICACCA)(Spring)*, Dehradun, India.

Kim, Yeongkwun, Injoo Kim, and Charlie Y Shim. 2014. "A taxonomy for DoS attacks in VANET." *2014 14th International Symposium on Communications and Information Technologies (ISCIT)*, Incheon, South Korea.

Kumar, Ankit, and Madhavi Sinha. 2014. "Overview on vehicular ad hoc network and its security issues." *2014 International conference on computing for sustainable global development (INDIACom)*, New Delhi, India.

Lipiński, Bartosz, Wojciech Mazurczyk, Krzysztof Szczypiorski, and Piotr Śmietanka. 2015. "Towards effective security framework for vehicular ad-hoc networks." *Journal of Advanced Computer Networks* 3 (2):134–140.

Malik, Vipin, and Savita Bishnoi. 2014. "Security threats in vanets: A review." *International Journal of Recent Research Aspects* 2:72–77.

Malla, Adil Mudasir, and Ravi Kant Sahu. 2013. "Security attacks with an effective solution for dos attacks in VANET." *International Journal of Computer Applications* 66 (22).: 45–49

Mejri, Mohamed Nidhal, Jalel Ben-Othman, and Mohamed Hamdi. 2014. "Survey on VANET security challenges and possible cryptographic solutions." *Vehicular Communications* 1 (2):53–66.

Mokhtar, Bassem, and Mohamed Azab. 2015. "Survey on security issues in vehicular ad hoc networks." *Alexandria Engineering Journal* 54 (4):1115–1126.

Nikam, AS Shahuraje, and Anshul Sarawagi. 2017. "Security over wormhole attack in VANET network system." *International Journal of Advanced Research Computer Science and Software Engineering* 7 (8).: 196–200

Nogueira, Michele, Helber Silva, Aldri Santos, and Guy Pujolle. 2012. "A security management architecture for supporting routing services on WANETs." *IEEE Transactions on Network and Service Management* 9 (2):156–168.

Nundloll, Vatsala, Gordon S Blair, and Paul Grace. 2009. "A component-based approach for (Re)-configurable routing in VANETs." *Proceedings of the 8th International Workshop on Adaptive and Reflective MIddleware*, Urbana Champaign Illinois.

Patil, Priyanka, Nilesh Marathe, and Vimla Jethani. 2016. "Improved ALERT protocol in MANET with strategies to prevent DoS & MITM attacks." *2016 International Conference on Automatic Control and Dynamic Optimization Techniques (ICACDOT)*, Pune, India.

Pooja, B, MM Manohara Pai, Radhika M Pai, Nabil Ajam, and Joseph Mouzna. 2014. "Mitigation of insider and outsider DoS attack against signature based authentication in VANETs." *2014 Asia-Pacific Conference on Computer Aided System Engineering (APCASE)*, South Kuta, Indonesia.

Qu, Fengzhong, Zhihui Wu, Fei-Yue Wang, and Woong Cho. 2015. "A security and privacy review of VANETs." *IEEE Transactions on Intelligent Transportation Systems* 16 (6):2985–2996.

Rawat, Ajay, Santosh Sharma, and Rama Sushil. 2012. "VANET: Security attacks and its possible solutions." *Journal of Information and Operations Management* 3 (1):301.

Ray, Partha Pratim. 2018. "A survey on Internet of Things architectures." *Journal of King Saud University-Computer and Information Sciences* 30 (3):291–319.

Sharma, Sparsh, and Ajay Kaul. 2018a. "Hybrid fuzzy multi-criteria decision making based multi cluster head dolphin swarm optimized IDS for VANET." *Vehicular Communications* 12:23–38.

Sharma, Sparsh, and Ajay Kaul. 2018b. "A survey on Intrusion Detection Systems and Honeypot based proactive security mechanisms in VANETs and VANET Cloud." *Vehicular Communications* 12:138–164.

Sharma, Surbhi, and Baijnath Kaushik. 2019. "A survey on internet of vehicles: Applications, security issues & solutions." *Vehicular Communications* 20:100182.

Singh, Amarpreet, and Priya Sharma. 2015. "A novel mechanism for detecting DoS attack in VANET using Enhanced Attacked Packet Detection Algorithm (EAPDA)." *2015 2nd international conference on recent advances in engineering & computational sciences (RAECS)*, Chandigarh, India.

Sumra, Irshad Ahmed, Iftikhar Ahmad, and Halabi Hasbullah. 2011a. Behavior of attacker and some new possible attacks in Vehicular Ad hoc Network (VANET). *2011 3rd International Congress on Ultra Modern Telecommunications and Control Systems and Workshops (ICUMT)*, Budapest, Hungary.

Sumra, Irshad Ahmed, Iftikhar Ahmad, and Halabi Hasbullah. 2011b. "Classes of attacks in VANET." *2011 Saudi International Electronics, Communications and Photonics Conference (SIECPC)*, Riyadh, Saudi Arabia.

Tangade, Shrikant S, and Sunilkumar S Manvi. 2013. "A survey on attacks, security and trust management solutions in VANETs." *2013 Fourth International Conference on Computing, Communications and Networking Technologies (ICCCNT)*, Tiruchengode, India.

Tayeb, Shahab, Matin Pirouz, Gabriel Esguerra, Kimiya Ghobadi, Jimson Huang, Robin Hill, Derwin Lawson, Stone Li, Tiffany Zhan, and Justin Zhan. 2017. "Securing the positioning signals of autonomous vehicles." *2017 IEEE International Conference on Big Data (Big Data)*, Boston, MA, USA.

Tobin, John, Christina Thorpe, and Liam Murphy. 2017. "An approach to mitigate black hole attacks on vehicular wireless networks." *2017 IEEE 85th Vehicular Technology Conference (VTC Spring)*, Sydney, NSW, Australia.

Verma, Swati, Bhawna Mallick, and Poonam Verma. 2015. "Impact of gray hole attack in VANET." *2015 1st International Conference on Next Generation Computing Technologies (NGCT)*, Dehradun, India.

Wu, Weigang, Zhiwei Yang, and Keqin Li. 2016. "Internet of vehicles and applications." In *Internet of Things*, 299–317. Morgan Kaufmann,Elsevier, USA.

Zaharuddin, Muhammad Hafiz Mazlan, Ruhani Ab Rahman, and Murizah Kassim. 2010. "Technical comparison analysis of encryption algorithm on site-to-site IPSec VPN." *2010 International Conference on Computer Applications and Industrial Electronics*, Kuala Lumpur, Malaysia.

Zeadally, Sherali, Ray Hunt, Yuh-Shyan Chen, Angela Irwin, and Aamir Hassan. 2012. "Vehicular ad hoc networks (VANETS): status, results, and challenges." *Telecommunication Systems* 50 (4):217–241.

Zhang, Tao, and Luca Delgrossi. 2012. *Vehicle Safety Communications: Protocols, Security, and Privacy*, Vol. 103: John Wiley & Sons, New Jeresy.

11

"Alexa, What about LGPD?": The Brazilian Data Protection Regulation in the Context of the Mediatization of Virtual Assistants

Fernando Nobre Cavalcante

CONTENTS

11.1 Introduction

In July 2019, in the United States, Mrs. Silvia Galva died during a fight with her boyfriend, Adam Crespo, having suffered an injury from a blade that pierced her chest. Debates about privacy and the right to information collected by virtual assistants heated new debates about data protection laws. Mrs. Galva's boyfriend is still the prime suspect, but politics still haven't been able to prove the crime. In its efforts to search for illuminating evidence, the policy requested a court order for the audio recordings made by Amazon Echo devices found at the scene of the crime. In response to the press, Amazon said the audio is not transmitted to the cloud or recorded unless it detects the certain specific words of: "Alexa", "Amazon" or "Echo". Once a recording is deleted, it cannot be recovered.

Five years before the crime, at the other edge of the Americas, Brazil took regulatory trials about the possibilities that the Internet of Things would bring to Brazilian citizens. Law 12 965 enacted on April 23, 2014, by President Dilma Rousseff, established rights and duties regarding access to information in Brazil. It is a rule framed in the Brazilian Constitution, that contributes to maintaining a "free and secure" Internet. However, it has not been exempt from being a point of continuous shocks and heated debates in the political sphere of Brazil. Known as the "Civil Internet Framework", the Law appeared along with 36 other projects of thematically similar proposals. It included controversial topics such as neutrality and citizen rights of access to the network. The first point of the "Civil Internet Framework" proposal refers to the relative neutrality of networks, which allows traffic control by private operators, seeking to maintain greater transparency with the user, expressed in Chapter 3, article 9. On the collection of user information, the framework guarantees, within a period of up to one year, the storage of connection data relating to navigation (IP address used, connection time, among others) by Internet providers. The objective, which in the background was to identify the user, would also protect him from cyber-crimes. of a "National Registry of Internet Access" (Nobre Cavalcante, 2017).

The "Internet Civil Framework" was the initial step toward a new law inspired by the European data protection model. The Brazilian Law 13.709/2018, entitled General Law on Personal Data Protection (LGPD) will require from any company two requirements: that the data be processed lawfully, and securely. This Law is inspired by the European Union's GDPR, (EU) 2016/679 and will be in force by August 2020, and will have a wide application, including for civil society organizations and public organs of the Union, States and Municipalities.

This study aims to approximate the field of study of data Sciences with media studies, emphasizing similar aspects among recent investigations on datafication in the face of the Internet of Things paradigm. Documentary analysis, using Atlas.ti software, will also be a contribution to instrumentalize the legislative aspects of European and Brazilian regulations raised here. Communication monopolies have found loopholes that make data privacy rights vulnerable. These points will be addressed in the methodological calls brought by this comparative study between two legislations. Gradually, the evidence of political mediatization in the reality of virtual assistants will be more notorious by debates about the regulations of them, signaling dominance of the Internet of Things (IoT) in social life. Currently, data protection laws provide evidence of that the legislative foundations for this are already laid. This study does not cover the aspects involving virtual assistants in their technical infrastructure constructions. However, it broadens horizons on the legal bases on which they are based.

Virtual assistants, like Alexa or Google Assistant, in the same way that they have had to comply with European data protection regulations, also find in LGPD barriers to the control of information that has been held by communication monopolies. Problematizing, mainly that LGPD finds in the political biases of mediatization important points for the understanding of the general laws of data protection, privacy and information security. In this section, the main concepts of datafication will be deepened, basing that these recent computational technologies of speed in the processing and in the increase of data storage that directly impact the new ways of handling and valuing information.

Although most contemporary researchers have prioritized the term 'mediatization' as a concept of a media-oriented multidimensionality process that produces meaning, few of them have attempted to detail how this dynamic has evolved, theoretically or empirically to the political field. By converging the logic of media and politics to the socio-discursive semiotic proposal, to the understandings of media studies, the research network entitled Communicative Figurations has launched important counterpoints to media studies in the field of politics (Hepp, 2011; Averbeck-Lietz, 2015). Assuming that political life is permeated by process of deep mediatization and communicable transmission intertwined, best theoretically illustrated by Nick Couldry, Andreas Hepp, and the research team at *Zentrum für Medien-, Kommunikations- und Informationsforschung* (ZeMKI) at the University of Bremen, they take a look at the questions about digital traits in the contemporary datafication process. In fact, digital traits can be understood from the relationship that a constellation of actors conceives by the rituality and vulnerabilities of their interactions in mediatized groups. Thus, the thematic frameworks are intertwined to mobilize a particular group. Combining IoT research with frames of relevance remains a challenge for researchers, even facing the open source communities that already undertake brackets analysis procedures of group interaction on Pyhton computational language (Cavalcante and Hanke, 2020).

According to Hepp et al., (2018), quoting Karanasios et al. (2013), the term 'datafication' refers to the increasing digitalization of media with software-based technology. For this reason, the authors update the media studies in the proposal to investigate the "digital tracks" of data that can be aggregated and processed in an automated way based on algorithms derived from the term. Big data technologies contextualize datafication in all social spheres, including interactions with voice assistants such as Alexa and Google Assistant. In the two sections following this chapter, the reader will find evidence of new actors who legislatively contribute to the datafication of voice assistance devices such as Alexa products; in particular, how this can be conflated with Brazilian and European data protection regulations.

11.2 LGDP vs GDPR: Comparisons between Brazilian and European Data Protection Regulations

Exchanges of information through virtual assistants, which are increasingly complex to construct, with deep mediatized reality, lack methodological and legal procedures to enhance national policies on data protection and privacy of citizens; especially when the theoretical context of the IoT enlightens these academic debates (Seo et al., 2017; Wachter, 2018). The regulations regarding how to investigate in the economic parameters on the stored and analyzed information of the users, hitherto held freely by the digital communication monopolies more recently, have to be submitted to the regulatory procedures of the European Union, made possible by the GDPR. There was also progress in Brazil with Law 13.709, popularly known as the General Data Protection Law (LGPD), sanctioned in 2018. The inspiration for this Brazilian Law took the main precepts of the European Union's GDPR (Mendes and Doneda, 2018). The chapter proposes to make a methodological overview drawing on documentary analysis of these two laws. Similarities were found between these laws, especially in the definitions of terms such as 'personal data', 'processing', 'data subject', 'pseudonymization', 'controller' and 'processor', stated in Article 4 of this European regulation.

The General Theory of State and Law is the keystone in this analysis between the field of political science and Law. This same field plan of politics is linked to natural Law. 'Natural', because it is related to the moral and ethical norms. It is a general conception that the Law imposes itself, not by the force of material coercion, but by the force of principles: universal and necessary moral norms, and that attends determined people. The politics action dimension is exactly when the common desire is projected. The social field regulated by politics can be the land of domination and oppression, or, through the conscious administration of the public welfare, the land of freedom and equity. Politics reflects the social interaction of people, which can generate conflicts of greater or lesser intensity.

Rescuing the historical aspects that help to understand the mediatization stages of the virtual assistants, it is necessary to highlight the inspirations of the legislative aspects that govern the protection of data of the National States. In classical philosophy, one of the pillars of natural Law is found in Aristotle's work *Politics*: the explanation of the 'Real State' based on the political organization of Athens and Sparta. Plato wrote about the State in *Res Publica* bringing the 'Ideal State,' as it should be in its conception of human beings and the world. In the sixteenth century, Machiavelli wrote *The Principatibus*, sowing the foundations of politics as an art of attainment; exercise and retention of power. This work has relevant value to the political world. *The Principatibus* is a doctrine addressed to public power.

Natural Law integrates the doctrine of *Jusnaturalismo*, that understands the existence of a system formed by conduct norms independent of the human will. St. Thomas Aquinas (1225–1274) glimpses three degrees in the hierarchy of laws: The Eternal Law, which is confused with the divine wisdom, The Natural, which governs the universe, and The Positive, emanated from the State. Positive Law embraces the world of the norms, which are mandatory rules of Law, that each person adopts to form their own Positive Law System. Hans Kelsen (1881–1973) brought to the legal world the classic work *Pure Theory of Law* (1934), the idea of the prevalence of the normative framework on the plan of 'Natural Law,' where exactly justice and politics are. Law should be completely unlinked from moral and ethical values, prevailing only the Law: seeking only purity. Law as legal science in excellence. In his foreword to the second edition, Kelsen admitted that he should include a chapter on natural and political Law. The history has shown that Law cannot be completely dissociated from politics; especially when the complexity of social relations deals with the datafication stages of the relationship between humans and non-humans (Kelsen, 1966).

The principles of the LGPD and GDPR are laid down the processing of personal data collected and processed, the duties of data processing agents, as well as the range of rights, entitle to the holder in the virtual environment. The full power of LGPD's personal data protection policies is based on Article 2 of this Legislation, according to the following parameters: (i) respect for privacy; (ii) informative self-determination; (iii) freedom of expression, information, communication and opinion; (iv) inviolability of intimacy, honor and image; (v) economic and technological development and innovation; (vii) free enterprise, free competition and consumer protection; and (vii) human rights, free development of personality, dignity and the exercise of citizenship by natural persons (LGPD 2018, art. 2). Even though the word 'privacy' is not mentioned in the GDPR, its meaning is clearly stated when Article 1 of that Law assumes "the protection of natural persons with regard to the processing of personal data" (GDPR 2018, art. 1).

Informative self-determination, expressed in Article 2 Section 2 of the LGPD, refers to efficient communication concerning to the collection of data, which acquires greater consistency in a scenario where consent is essential to trigger treatment, while the 'vice' of consent blocks exactly the possibility of personal operating data. Informational self-determination is also reflected in Article 5 of the LGPD as the subject's consent is "a free, knowledgeable and unequivocal manifestation whereby the data subject agrees to the processing of his/her personal data for a particular purpose" (LGPD, 2018, art. 5, translated). The GDPR devotes four sections in article 7 to the conditions applicable to consent, as follows: "the request for consent shall be presented in a manner which is clearly distinguishable from the other matters, in an intelligible and easily accessible form, using clear and plain language" (GDPR 2018, art. 7). In both laws, the data subject is considered vulnerable to a possible lack of understanding about the consent information; for this reason, the information about data collection and processing must be expressly clear. A full withdrawal and temporary suspension of any data processing operation are guaranteed by both laws as well as ensuring freedom of expression, protection of human rights, including regulations on the collection and processing of children's data. Likewise, in both laws, the economic development, the rights of states, free competition, and the consumer's right are also guaranteed.

In three sections, Article 91 of the GDPR sets out data protection standards for churches and religious associations. On the other hand, no reference is reported in LGPD. IoT provides a conceptualization that drives various design techniques to achieve different efficiency and high-performance goals (Gregorio et al. 2020). The regulations that impose the companies to informative self-determination standards, allows the user a greater understanding of the data processing on their communicative actions of the ultra-connected devices in this type of infrastructure. "What is of more significance is that the IoT demonstrates how digitization was not associated with the convergence of all digital media into one ultimate device," (Hepp 2019: 42). In the datafication process, devices have increasingly become ways of communication that direct users to interconnected daily practices. A critical understanding of how data is processed, paying attention to the detail on the processed data without falling into the habit of accepting the terms without reading them, is a barrier to the new data protection regulations.

If public policies in large cosmopolitan cities are already adapting to the reality of IoT's infrastructural complexities (Grisot et al., 2018), understanding how these data protection policies complement each other becomes urgent. The legal structure of LGPD and GDPR follows similarity of topics, explicit in the legal chapters, as presented in Table 11.1. The LGPD makes international data transfer processes clearer in Chapter V and details the creation of a responsible body through the National Data Protection Authority (ANPD) in Chapter IX of this act. A similar approach is advocated in the GDPR, especially, in Section 3, to detail the role of The European Data Protection Board ('the Board'). However, the methodological design in cooperative work, the tasks of the Board, the rules of mandates are better detailed in GDPR.

TABLE 11.1

Comparison of Legal Text Structure. (Source: author's elaboration).

Comparison Perspectives	GDPR	LGDP
Number of Legal Articles	99 articles	65 articles
Sections of Legal article chapters	Chapter I: *General provisions* Chapter II: *Principles relating to processing of personal data* Chapter III: *Rights of the data subject* Chapter IV: *Controller and processor* Chapter V: *Transfers of personal data to third countries or international organizations* Chapter VI: *Independent supervisory authorities* Chapter VII: *Cooperation and consistency* Chapter VIII: *Remedies, liability and penalties* Chapter IX: *Provisions relating to specific processing situations* Chapter X: *Delegated acts and implementing acts* Chapter XI: *Final provisions*	Chapter I: *Preliminary provisions* Chapter II: *Requirements relating the processing of personal data* Chapter III: *Rights of the data subject* Chapter IV: *Processing of personal data by the public authorities* Chapter V: *International data transfer* Chapter VI: *Personal data processing agents* Chapter VII: *Security and good practices* Chapter VIII: *Judicial Review* [Surveillance] Chapter IX: *National Data Protection Authority (ANPD) and the national council for the protection of personal data and privacy* Chapter X: *Final and Transitory Provisions*

The legal aspects for the handling of agents' data by controllers and processors are addressed in Chapter 2 of LGPD. Requirements, terminations, 'sensitive data' (*dados sensíveis*, in Portuguese), and data from children and adolescents, are explained in four sections of this chapter. Taking a similar perspective, the GDPR defines in Article 4 typologies that the Law deals with as well as the data processing agents. Article 5 of the LGPD defines personal data as "information relating to an identified or perceptible person"; sensitive data as "personal data concerning racial or ethnic origin, religious conviction, political opinion, trade union membership or religious, philosophical or political organization, data concerning health or sex life, and genetic or biometric data when linked to a natural person," (LGPD 2018, art. 2, translated). The nomenclature 'sensitive data' is not used by the GDPR. However, the meaning that the LGPD applies to this concept is found in the definitions of 'genetic data', 'biometric data' and 'data concerning health', as described in Article 4 of the GDPR. It can also be found in the definition of 'personal data': "an identifiable natural person is one who can be identified, directly or indirectly, … specific to the physical, physiological, genetic, mental, economic, cultural or social identity of that natural person," (GDPR 2018, art. 4). In the GDPR, the legal conditions and aspects on child data are contained in articles 8, 9, 12, 40 and 57. In both laws the language on a child's data processing should be as simple and clear as possible because of the potential for greater vulnerabilities in the web, as well as agreements with the consent of the holder of the child's parental responsibilities. According to what is explained in subsection 6 of Article 14 of the LGPD, data regarding children and adolescents must foresee the physical-motor, perceptual, sensory, intellectual and mental characteristics of the user, using audiovisual resources when appropriate, to provide the necessary information to the parents or legal guardian, and appropriate for the child's understanding.

The principles of the GDPR, which are similar to those of the LGPD above cited, are set out in Chapter I of this Law, and also ratified in Chapter III which expresses details of the rights of the data subject, and Chapter IV is dedicated to explaining the rights and duties of controllers and processors. The structure and the ways listed by the business actors involved in data processing, the transfers of personal data and their purposes, the legal character and the authorities involved in the European Union, the rights of data subjects, penalties for data breach and how communication should be passed on to users, are clearly in the GDPR, being better organized than in the LGPD. Often, information on a given topic is grouped in blocks in this Law. Each article in the GDPR presents a title that synthesizes it, differently from the structure in the LGPD.

In this documentary analysis[1], the term 'personal data' (*dados pessoais* in Portuguese) was found 149 times in the ten chapters of the LGPD. The term is associated with contents that explain the data processing and the obligations of the controllers and processors of the data subjects' information. The first comparison of this term between Brazilian and European laws is made in Table 11.2. Possibly, because the Portuguese language uses more relative pronouns and subject and object pronouns, the term 'personal data' was found less frequently in the LGDP, as expressed in Table 11.2. Words in Portuguese such as *'isso'*, *'esse'* or oblique pronouns like *'lo'*, *'la'*, *'o'*, *'a'* were not considered in this documentary research.

In Chapter IV of the GDPR, the term 'personal data' is more explicit, referring to those responsible for processing and subcontracting user data. The same narrative strategy occurs in the LGPD, where the term 'personal data' is commonly associated with data processing practices in Chapter II. It is important to note that Hepp (2019) and Hintz et al. (2019) have already demonstrated how datafication studies can help to understand the importance of GDPR on citizens' data.

TABLE 11.2

Comparison of the Term 'Personal Data'. (Source: Author's Elaboration).

Comparison Perspectives	GDPR	LGDP
Translation of *"Personal data"*, *"dados pessoais"* and *"dado pessoal"* in Portuguese without reference pronouns.	254 occurrences under ten distinct chapters: **Chapter IV:** Fourteen articles (Arts. 25, 27–30, 32–35, 37–40 and 42) **Chapter III:** Eleven articles (Arts. 13–23) **Chapter II:** Seven articles (Arts. 5–11) **Chapter V:** Seven articles (Arts. 44–50) **Chapter I:** Four articles (Arts. 1–4) **Chapter VI:** Four articles (Arts. 51, 53, 57 and 58) **Chapter VIII:** Four articles (Arts. 77, 79, 80 and 83) **Chapter IX:** Four articles (Arts. 86, 88–90) **Chapter XI:** Two articles (Arts. 94 and 97) **Chapter VII:** One article (Art. 70)	149 occurrences under ten distinct chapters: **Chapter II:** Nine articles (Arts. 7–9, 11–16) **Chapter I:** Six articles (Arts. 1–6) **Chapter IV:** Six articles (Arts. 26, 27, 29–32) **Chapter VII:** Six articles (Arts. 46–51) **Chapter III:** Five articles (Arts. 17–21) **Chapter VI:** Five articles (Arts. 41–45) **Chapter V:** Four articles (Arts. 33, 34, 37 and 38) **Chapter IX:** Four articles (Arts. 55-J, 55-K, 58-A and 58-B) **Chapter VIII:** One article (Art. 52) **Chapter X:** One article (Art. 60)

> The regulation of digital media and their infrastructures are not simply about the protection of individual rights. As important as the General Data Protection Regulation (GDPR) act in the European Union is for the extensive protection of citizens' personal data (Hintz et al., 2019: 68–74), regulation should also enable other forms of organization and a *Gestaltung* of deep mediatization to a greater degree.
>
> **(Hepp, 2019:190)**

Article 8 of the LGPD details how personal data should be authorized. The holder's consent shall be given as follows: if consent is given in writing, it must be included in a separate clause of the other contractual clauses; the processing of personal data through consent is prohibited; consent shall refer to specified purposes, and generic authorizations for the processing of personal data shall be null and void. It is important to ratify that consent may be revoked at any time through an express statement by the holder, employing free and facilitated procedure, and those treatments carried out under the protection of the previously expressed consent be ratified as long as there is no request for elimination. Similar conditions of the LGPD are set out in Article 7 of the GDPR.

The legal aspects of the LGPD and the GDPR data processing imply any processing operation carried out by a natural person or by a public or private legal entity, being these respectively Brazilian or from any country of the European Union. It does not only refer to data processing agents, but especially if the processing actions reach at least one citizen of these countries. In the case of the LGPD, the Law does not apply to the following cases regarding data processing:

(i) data carried out by a natural person for exclusively private and non-economic purposes;

(ii) data carried out exclusively for:
a. journalistic and artistic purposes; or
b. academic purposes;

(iii)   data carried out exclusively for:
    a.   public security;
    b.   national defense;
    c.   state security; or
    d.   activities of investigation and repression of criminal infractions coming from outside the national territory.

(LGPD 2018, art. 5)

The GDPR does not concern issues of defense of fundamental rights and freedoms or the free transit of personal data relating to activities which fall outside the scope of the Union law, such as those relating to national security. The regulation also does not apply to the processing of personal data by the Member States when carrying out activities relating to the Union's common foreign and security policy. Like the LGPD, the GDPR does not apply when the processing of personal data is carried out by individuals in the exercise of exclusively personal or domestic activities, or for purposes of personal freedom of expression such as journalistic, artistic and academic activities. It is more explicit in Article 85 of the GDPR:

> For processing carried out for journalistic purposes or the purpose of academic artistic or literary expression, Member States shall provide for exemptions or derogations from Chapter II (principles), Chapter III (rights of the data subject), Chapter IV (controller and processor), Chapter V (transfer of personal data to third countries or international organizations), Chapter VI (independent supervisory authorities), Chapter VII (cooperation and consistency) and Chapter IX (specific data-processing situations) if they are necessary to reconcile the right to the protection of personal data with the freedom of expression and information.

(GDPR 2018, art. 85)

In this Article 5 of the LGPD, data processing is defined as any operation carried out with personal data, such as those relating to "the collection, production, reception, classification, use, access, reproduction, transmission, distribution, processing, archiving, storage, deletion, evaluation or control of information, modification, communication, transfer, dissemination or extraction" (LGPD 2018, art. 5; translated). A similar definition is found in Article 5 (clause 2) of the GDPR:

> 'processing' means any operation or set of operations which is performed on personal data or on sets of personal data, whether or not by automated means, such as collection, recording, organization, structuring, storage, adaptation or alteration, retrieval, consultation, use, disclosure by transmission, dissemination or otherwise making available, alignment or combination, restriction, erasure or destruction.

(GDPR 2018, art. 5)

The criteria for the processing of data in the GDPR must follow at least one of these conditions:

(i)   the consent to the processing of personal data for one or more specified purposes;

(ii)   the signing of a contract in which the data subject is a party;

(iii)   compliance with a legal obligation;

(iv)   the protection of the vital interests of the data subject;

TABLE 11.3

Comparison of the Term 'Processing.' (Source: Author's Elaboration).

Comparison Perspectives	GDPR	LGDP
Translation of "Processing" in Portuguese as *"tratamento de dados"* without reference pronouns.	304 occurrences under ten distinct chapters: **Chapter IV:** Fifteen articles (Arts. 24, 26– 30, 32, 35–42) **Chapter III:** Twelve articles (Arts, 12–23) **Chapter II:** Seven articles (Arts. 5–11) **Chapter I:** Six articles (Arts. 1–6) **Chapter VIII:** Six articles (Arts. 77, 79– 83) **Chapter IX:** Six articles (Arts. 85– 91) **Chapter VI:** Five articles (Arts. 51, 55–58) **Chapter VII:** Three articles (Arts. 62, 64 and 71) **Chapter XI:** Three articles (Arts. 94, 95 and 98) **Chapter V:** Two articles (Arts. 44 and 47)	64 occurrences under eight distinct chapters: **Chapter II:** Eight articles (Arts. 7– 11, 14–16) **Chapter I:** Six articles (Arts. 1–6) **Chapter VI:** Four articles (Arts. 41, 42, 43 and 44) **Chapter IV:** Two articles (Arts. 23 and 31) **Chapter V:** Two articles (Arts. 37 and 38) **Chapter VII:** Two articles (Arts. 49 and 50) **Chapter IX:** Two articles (Arts. 55-J and 58-A) **Chapter VIII**: One article (Art. 52)

(v) the exercise of functions carried out in the public interest or the use of official authority;

(vi) the necessity of the legitimate interests pursued by the controller or by a third party, and which does not apply to the processing carried out by public authorities.

(GDPR 2018, art. 6)

Chapter II of this Law is dedicated to the legal principles of data processing and their regulatory applicability. Also, Chapter IV provides an extensive definition of the functions, processes and conformities that controllers and processors are subject. This comprehensive specification of the GDPR makes it more detailed than the LGPD in respect of data processing agents. It can be better observed in the comparative box of Table 11.3, which shows the higher frequency of the expression 'data processing' in the GDPR.

By processing data that impute possibilities of risks, on personal privacy and freedom to data subjects, the LGDP and GDPR underline the rights of these subjects to be informed of the nature, scope, context and purposes of their data processing. The risk should be evaluated based on an objective assessment, by which the degree of risk of the data processing operations is established. According to Article 5 of the LGPD, the data subject is "the natural person to whom the personal data being processed refers" (LGPD, 2018, art. 5; translated). Such a definition is given in Article 4 of the GDPR: "relating to an identified or identifiable natural person ('data subject'); an identifiable natural person is one who can be identified, directly or indirectly". (GDPR 2018, art. 4). In both laws, the frequency of "data subject" terms are more associated with the processing of their data by controllers and processors. As shown in Table 11.4, this association can be found in Chapter IV of GDPR and Chapter II of LGPD. Although LGPD has a lower frequency of the term 'data subject,' both laws address similar precautions about the processing of data subjects. However, it is noted that the GDPR concentrate more detailed information on the responsibilities of controllers and processors. Similarly, the LGPD better highlight issues about sensitive data concerning data subjects.

TABLE 11.4

Comparison of the Term 'Data Subject'

Comparison Perspectives	GDPR	LGDP
Translation of "Data subject" in Portuguese as *"titular dos dados"* and *"titulares dos dados"*, without reference pronouns.	209 occurrences under nine distinct chapters: **Chapter IV:** Fifteen articles (Arts. 25–28, 30, 33–43) **Chapter III:** Twelve articles (Arts. 12–23) **Chapter II:** Six articles (Arts. 5–7, 9–11) **Chapter VII:** Five articles (Arts. 60, 62, 65, 66 and 70) **Chapter VIII:** Five articles (Arts. 77–80 and 83) **Chapter V:** Four articles (Arts. 45–47 and 49) **Chapter VI:** Three articles (Arts. 56–58) **Chapter IX:** Three articles (Arts. 87–89) **Chapter I:** Two articles (Arts. 3 and 4)	97 occurrences under ten distinct chapters: **Chapter II:** Seven articles (Arts. 7–11, 14 and 15) **Chapter VII:** Six articles (Arts. 46–51) **Chapter III:** Five articles (Arts. 17–21) **Chapter VI:** Five articles (Arts. 41–45) **Chapter IV:** Three articles (Arts. 23, 26 and 27) **Chapter V:** Three articles (Arts. 33, 35 and 36) **Chapter I:** Three articles (Arts. 4–6) **Chapter VIII:** One article (Art. 52) **Chapter IX:** One article (Art. 55-J) **Chapter X:** One article (Art. 60)

Source: Author's elaboration.

Both Chapter IV of the GDPR and Chapter VII of the LGPD guarantee the data processing security policies of them, even in the event of information loss or misuse, the subjects must be notified. Many peculiarities make the laws similar, covering certain points more or less. As this study does not make a hermeneutic analysis of legal objects, it signals the need for further investigations to scrutinize the common and the opposite aspects of these laws.

Anonymized data as a "data relating to a holder who cannot be identified, considering the use of reasonable and available technical means at the time of their processing" is defined in Article 5 of the LGPD. In this same article, anonymization or pseudonymization, is considered as the use of reasonable and available technical resources at the time of treatment, whereby the possibility an association, directly or indirectly, to an individual is lost. In Brazilian Law, inspired in the GDPR, the pseudonymization process ensures that processors and data controllers keep the identity of the subjects confidential.

> 'pseudonymisation' means the processing of personal data in such a manner that the personal data can no longer be attributed to a specific data subject without the use of additional information, provided that such additional information is kept separately and is subject to technical and organisational measures to ensure that the personal data are not attributed to an identified or identifiable natural person.
>
> **(GDPR, 2018 art. 4)**

In both laws, the protection of individuals' rights and freedoms concerning to the processing of personal data requires that appropriate technical and organizational measures be taken to ensure internal policies to implement measures that respect data protection principles by making the process transparent and preferably followed by pseudonymization methodologies. As much as the GDPR and the LGPD do not explicitly cite data processing over the IoT, this is implicit in the responsibilities of data processing agents with advanced technologies using pseudonym data. The LGPD is redundant when it mentions that

TABLE 11.5

Comparison of the Term 'Pseudonymization' (Source: Author's Elaboration).

Comparison Perspectives	GDPR	LGDP
Translation of "Pseudonymization" in Portuguese as *"pseudonimização"* *"anonimação".*	6 occurrences under four distinct chapters. **Chapter I:** One article (Art. 4) **Chapter II:** One article (Art. 6) **Chapter IV:** Three articles (Art. 25, 32 and 40) **Chapter IX:** One article (Art. 89)	16 occurrences under three distinct chapters: **Chapter II:** Four articles (Arts. 7, 12, 13 and 16) **Chapter I:** One article (Art. 5) **Chapter III:** One article (Art. 18)

research centers should seek pseudonymization practices, especially in Chapter II. It can be seen in Table 11.5 which shows the highest incidence of the term in Chapter II.

There is a distinction between 'controllers' and 'processor' in both laws. The LGPD in Article 5 defines a controller as "a natural or legal person, governed by public or private law, who is responsible for decisions concerning the processing of personal data," (LGPD 2018, art. 5; translated). In the same article, the processor is a "natural or legal person, whether governed by public or private law, who carries out the processing of personal data on behalf of the controller," (LGPD 2018, art. 5; translated). The GDPR has a similar definition for these two agents, expressed in Article 4 of the Regulation: "'processor' means a natural or legal person, public authority, agency or other body which processes personal data on behalf of the controller" (GDPR 2018, art. 4) and "'controller' means the natural or legal person, public authority, agency or other body which, alone or jointly with others, determines the purposes and means of the processing of personal data" (GDPR 2018, art. 4). These terms are compared between the two regulations in Tables 11.6 and 11.7.

TABLE 11.6

Comparison of the Term 'Controller'

Comparison Perspectives	GDPR	LGDP
Translation of "Controller" in Portuguese as *"controlador"* and *"controladores"* without reference pronouns	307 occurrences under nine distinct chapters: **Chapter IV:** Twenty articles (Arts. 24–43) **Chapter III:** Twelve articles (Arts. 12– 23) **Chapter II:** Five articles (Arts. 5– 9) **Chapter V:** Five articles (Arts. 44, 46–49) **Chapter VI:** Four articles (Arts. 56–59) **Chapter VII:** Four articles (Arts. 60, 62, 65 and 70) **Chapter VIII:** Four articles (Arts. 79, 81–83) **Chapter I:** Two articles (Arts. 3 and 4) **Chapter IX:** Two articles (Arts. 85 and 90)	62 occurrences under eight distinct chapters: **Chapter II:** Eight articles (Arts. 7– 11, 13, 14 and 16) **Chapter V:** Four articles (Arts. 33, 37–39) **Chapter VI:** Three articles (Arts. 41, 42 and 44) **Chapter III:** Two articles (Arts. 18 and 20) **Chapter VII:** Two articles (Arts. 48 and 50) **Chapter I:** One article (Art. 5) **Chapter VIII:** One article (Art. 52) **Chapter IX:** One article (Art. 55J)

Source: Author's elaboration.

TABLE 11.7

Comparison of the Term 'Processor'.

Comparison Perspectives	GDPR	LGDP
Translation of "Processor" in Portuguese as *"operador"* without reference pronouns	174 occurrences under eight distinct chapters: **Chapter IV:** Seventeen (Arts. 27–44) **Chapter V:** Four articles (Arts. 46–49) **Chapter VII:** Four articles (Arts. 60, 62, 65 and 70) **Chapter VIII:** Four articles (Arts. 79, 81–83) **Chapter VI:** Three articles (Arts. 56–58) **Chapter I:** Two articles (Arts. 3 and 4) **Chapter IX:** Two articles (Arts. 85 and 90) **Chapter III:** One article (Art. 23)	14 occurrences under four distinct chapters: **Chapter V:** Three articles (Arts. 35, 37 and 39) **Chapter VI:** Two articles (Arts. 42 and 44) **Chapter VII:** One article (Art. 50) **Chapter I:** One article (Art. 5)

Source: Author's elaboration.

In both cases, the incidence of the terms compared is higher in the GDPR than in the LGPD. Although the processors do not have direct contact with the data subjects, since they are the ones who operate the data processed from the users, it is required by both laws that they have the same responsibility as those who control these data. Article 26 of the GDPR makes this co-responsibility explicit: "Where two or more controllers jointly determine the purposes and means of processing, they shall be joint controllers," (GDPR 2018, art. 26). The same is guaranteed in Article 42 of the LGPD. Any subcontractor under contract with the processors must have the explicit consent of the data controllers and the object and duration of the processing, the nature and purpose of the processing, the type of personal data and categories of data subjects, and the obligations and rights of the controller. Helberger et al. (2018), Cavalcante (2019) and Hepp (2019) highlight the look of datafication studies at aspects of co-responsibility among these agents in GDPR.

> The principle of 'cooperative responsibility' (Helberger et al., 2018: 1) is fundamental here and needs to be considered when questioning platform regulation. With platforms key public values such as transparency, diversity and civility can only be secured through the involvement of various stakeholders such as the companies that operate the platform, its users and public institutions.
>
> **(Hepp, 2019:190)**

Chart 1, as well as the tables presented, summarizes the documental analysis of the terms assessed in this study. The most considerable disproportion of the terms compared are 'processor', 'controller' and 'processing' respectively. What can be gauged so far in relation to the three disproportional terms presented in Chart 1, is that these are more detailed in the GDPR than in the LGPD. This may open loopholes for possible legal proceedings against the data processing and its responsible agents (Figure 11.1).

This study, as previously mentioned, did not aim to make a hermeneutic analysis; that means neither rhetorical nor interpretative arguments are exploited. However, the documental analysis will allow future studies to explore better the conflicting and similar ideas of these two laws.

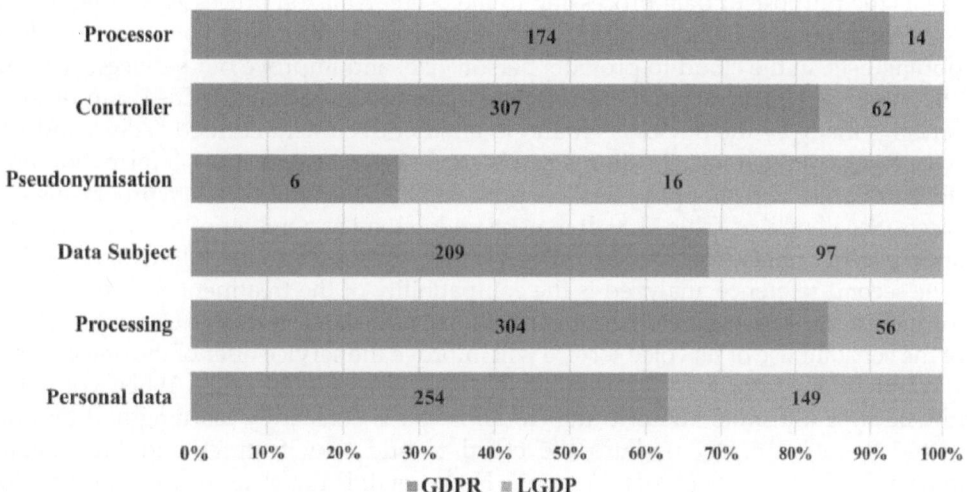

FIGURE 11.1
Proportional comparison between the terms sought in the documentary search. (Source: Author's elaboration).

11.3 Brazilian Data Protection Regulations in the Context of Alexa Echo Dot's Privacy Policy

Recent studies have already dedicated to understanding the legal aspects that GDPR gives to the voice assistant scenario. (Flint, 2017; Furey and Blue, 2018a; Furey and Blue, 2018b; Ni Loideain and Adams, 2018). However, there is an academic shortage that focuses on the reality of the Brazilian regulations on virtual assistants, especially Alexa products. This section of the chapter contrasts the LGPD's articles on data protection and privacy with the Alexa Echo Dot's Terms of Use updated on October 30, 2019. To this end, this section aims to contrast the ten principles of LGPD, which are based on the GDPR, with the Terms of Use (*Termos de Uso da Alexa*, in Portuguese), Conditions of Use (*Condições de Uso*, in Portuguese) and Alexa's FAQ (*Perguntas Frequentes sobre a Alexa e os Dispositivos Alexa*, in Portuguese). The documents analyzed are in Portuguese and accessible on the Internet. For the translation of some quotations from this text, the English version of the same document was used in the original to preserve the corporate tone of the company version.

Before the LGPD, the Access to Information Law (*Lei de Acesso à Informação*) in 2011; the Computer Crimes Law (*Lei de Crimes Informáticos*) in 2012, and the Internet Civil Framework (*Marco Civil da Internet*) in 2014 ordered the Internet in Brazil as a jurisdiction environment. In its initial proposal, the LGPD comes into force in August 2020, but a recent bill (5762/19) by the House of Representatives intends to postpone it to 2022. One of the reasons is the delay[2] in the creation and organization of the National Data Protection Authority (ANPD).

The first analytical axis that this documentary analysis sought to make refers to if the "legitimate, specific, explicit and informed purposes of the subject" are explicit in Alexa's Terms of Use and Privacy. In item 1.3 of Alexa's Terms of Use, the company minimally presents the purpose of data processing to the user: "Amazon processes and retains your Alexa Interactions, such as your voice inputs, music playlists, and your Alexa to-do and shopping lists, in the cloud to provide, personalize, and improve our services," (Amazon 2019, *Termos de Uso da Alexa* 1.3). Complete information on how the user can delete the recordings made by the devices is found in another document entitled "Alexa and Alexa Device FAQs". In general, the Terms of Use reviewed presents more information on the usability of voice and shopping services provided by Alexa than on data processing of the subjects (See item 'i' of Table 11.8). It may be understood as a matter of prevention of information security policies for the subjects, as well as illustrated in the following.

The second reference analyzed is the compatibility of the treatment with the purposes informed to the subjects (See item 'ii' of Table 11.8). Alexa's Terms of Use expressly state that the constant use of its voice service will improve the service offer of the device: "Alexa is a continuously improved service that you control with your voice. When you interact with Alexa, it transmits audio to the cloud. Alexa is constantly learning and becoming smarter. It automatically updates the cloud to add new features and functionality," (Amazon 2019, *Termos de Uso da Alexa* 1.3). However, it is not clear in any company document how these services are improved. The treatment agents involved in setting up the roles of the processors are not presented in the Terms of Use.

The limitation of data processing to the minimum necessary for the realization of its purposes, principle of LGPD, is implicit in the Terms of Use and more clearly in the document "Alexa and Alexa Device FAQs". There is no explicit evidence in the Alexa's Terms of Use document that the data management of Alexa's services makes minimal use of the users' data for information processing (See item 'iii' of Table 11.8). However, this information can be found in Section 4 of the "Alexa and Alexa Device FAQs" when this question is asked: "4. How does Alexa minimize the amount of data sent to the cloud?". The answer to this in the document is "Alexa and Echo devices are designed to record as little audio as possible and minimize the amount of background noise streamed to the cloud." It can be regarded as a key item for company information security policies, as expressed in the following analyses.

The fourth point analyzed concerns the free and easy consultation on the form and duration of data processing using Alexa products. Items 3.3 and 3.5 of Alexa's Terms of Use document remove any kind of liability from the company, making unclear the period and limits of data processing. "We may change, suspend, or discontinue Alexa, or any part of it, at any time without notice." (Amazon 2019, *Termos de Uso da Alexa* 3.3). "Your rights under this Agreement will automatically terminate without notice if you fail to comply with any of its terms. In case of such termination, Amazon may immediately revoke your access to Alexa without refund of any fees," (Amazon 2019, *Termos de Uso da Alexa* 3.5). As presented in the previous section of this study, the conditions imposed by Alexa's Terms of Use do not respect the determinations of free access of LGPD information (See item 'iv' of Table 11.8.

The LGPD guarantees to the Brazilian citizens' accuracy, clarity, relevance and updating of data on user information processing. This fifth point raised in this investigation is not tangible in Alexa's Terms of Use (See item 'v' of Table 11.8). Item 3.2, contrary to what should be complied with by the Brazilian regulations, attributes to the user any blame in the eventual services distorted or not provided. "We do not guarantee that Alexa or its functionality or content (including traffic, health, or stock information) is accurate, reliable, always available, or complete. You may encounter content through Alexa that you find offensive, indecent, or objectionable," (Amazon 2019, *Termos de Uso da Alexa* 3.2).

TABLE 11.8

Comparison of Alexa's Terms of Use in the Portuguese Version of 10/30/19 with the Ten Principles of LGPD

LGPD	Alexa's Terms of Use
(i) purpose: carrying out the processing for legitimate, specific, explicit and informed purposes to the subject, without any possibility of further processing incompatible with those purposes; (precise in Section I of Article 6)	The purpose of the processing is not so exact in Alexa's Terms of Use document in Portuguese. Separate information is found in a second document entitled "Alexa and Alexa Device FAQs". **Evidence**: Item 1.3.
(ii) appropriateness: compatibility of the treatment with the purposes informed to the data subject, according to the context of the procedure; (explicit in Section II of Article 6)	The use of the service provision seems to align the need for data processing; however, it is not expressed how information processing happens. **Evidence**: Item 1.3.
(iii) necessity: limitation of processing to the minimum necessary for the processing, with a scope of the relevant data, proportionate and not excessive about the purposes of the processing; (explicit in Section III of Article 6)	No unambiguous evidence was found in Alexa's Terms of Use. This is set out in the FAQ section of a separate document. **Evidence**: "Alexa and Alexa Device FAQs", question number 4.
(iv) free access: guarantee to data subjects easy and free consultation of the form and duration of the processing as well as the completeness of their personal data; (explicit in Section IV of Article 6)	There is disagreement with the LGPD principle of free access to information. **Evidence**: Items 3.3 and 3.5.
(v) data quality: guarantee to the data owners, of accuracy, clarity, relevance and updating of the data, according to the need and for the fulfillment of the purpose of their processing; (explicit in Section V of Article 6)	The quality of information processing is not guaranteed by Alexa's Terms of Use. **Evidence**: Item 3.2.
(vi) transparency: the data subjects are guaranteed clear, precise, and easily accessible information on the processing carried out and on the relevant processing agents, with due regard for commercial and industrial secrets; (explicit in Section VI of Article 6)	Alexa's Terms of Use do not present transparent information about the user's data processing. **Evidence**: "Alexa Communication Schedule".
(vii) security: use of technical and administrative measures to protect personal data from unauthorized access and accidental or unlawful destruction, loss, alteration, communication or dissemination; (explicit in Section VII of Article 6)	Alexa's Terms of Use and the FAQ does not provide any information on data processing security. **Evidence**: No evidence.
(viii) prevention: adoption of measures to prevent the occurrence of damage due to the processing of personal data; (explicit in Section VIII of Article 6)	Alexa's Terms of Use and the FAQ does not provide any information on data processing prevention. **Evidence**: No evidence.
(ix) non-discrimination: inability to carry out treatment for unlawful or abusive discriminatory purposes; (explicit in Section IX of Article 6)	In the Conditions of Use section that Alexa's Terms of Use refers to there is evidence only about the prohibition of purchases of its products and services for persons under the age of 18. **Evidence**: "Conditions of Use section."
(x) liability and accountability: demonstration, by the agent, of the adoption of effective measures capable of proving compliance and enforcement with personal data protection regulations, including the effectiveness of such actions. (explicit in Section X of Article 6)	Alexa's Terms of Use and the FAQ does not provide any information on data processing compliance. **Evidence**: No evidence.

Source: Author's elaboration.

In compliance with trade and industry secrets, LGPD guarantees users of technological services, including Alexa products, information that must be clear, precise and easily accessible about the processing and the respective data processing agents. Certain Alexa services are interlinked with its network of partners, and this is expressed in the annex entitled "Alexa Communication Schedule" (*"Anexo sobre o Alexa Communication"*, in Portuguese). AMCS LLC ("AMCS"), an affiliate of Amazon, however, does not guarantee to the user any liability for the products of its partner companies. It is expressed in the following message in the initial part of the Annex: "AMCS and its affiliates may offer services other than Alexa Communication, which are not covered by this schedule and may be subject to other terms." The "Other Terms" are not contained in Alexa's Portuguese version from October 30, 2019. Even if business secrets do not need to be revealed to users, the company does not even provide additional links about the "other terms" mentioned. Especially and even though they are fundamental to subjects' information processing. Alexa's Terms of Use in Portuguese version, to which this article refers the translation of the original in English, does not seem to follow the transparency required by LGPD (See item 'vii' of Table 11.8). At least until this moment.

The seventh principle of the LGPD contrasted here refers to "the use of technical and administrative measures to protect personal data from unauthorized access and accidental or unlawful destruction, loss, alteration, communication or dissemination," (LGPD 2018, art. 6). There is no reference to the term data security, or any other relationship to information security, in Alexa's Terms of Use (See item 'vii' of Table 11.8). In the FAQ section, which is an editable link for the company, there is only this mention: "We have a team of world-class scientists and engineers dedicated to continually improving our wake word detection technology and preventing false wakes from happening, including through the cloud verification mechanism," (Alexa 2019, *Perguntas Frequentes*). Besides, this study did not find any information regarding the company's data processing compliance precautions. No reference to this is made in Alexa's Terms of Use or the FAQ sections (See item 'viii' of Table 11.8). The ninth principle guaranteed by the LGPD refers to the impossibility of carrying out data processing for discriminatory purposes. No information about the race, gender, sexual orientation or religion is mentioned in Alexa's Conditions of Use section (See item 'ix' of Table 11.8). However, the company makes it clear in the FAQ section that no person under the age of 18 may purchase its products, except with a legal representative. Non-discrimination policies could be clarified in the documents analyzed.

The last principle of the LGPD refers to the adoption of effective measures capable of proving compliance with and enforcement of personal data protection regulations (See item 'x' of Table 11.8). The company does not present a clear data processing compliance policy communicated to users in plainly and simply manner as expressed in LGPD. As expressed by the company, "Your rights under this Agreement will automatically terminate without notice if you fail to comply with any of its terms." And "In case of such termination, Amazon may immediately revoke your access to Alexa without refund of any fees. Amazon's failure to insist upon or enforce your strict compliance with this Agreement will not constitute a waiver of any of its rights," (Alexa 2019, *Termos de Uso da Alexa* 3.5).

After observing this analysis, it is ratified that Alexa's Terms of Use do not fully comply with the LGPD. There has been little exploration of data processing procedures, pseudonymization and the rights and obligations of the processors involved in the user data processing. The fact that LGPD has not yet entered into force in Brazil until the completion of this study implies that there is still much to be done by companies in terms of information security datafied by the IoT technologies.

11.4 Conclusion

Contemporary policies have the legitimacy (or not) to make institutional forces capable of deepening the current massification of mediatized devices in all social areas. Signaling a range of definitions about the waves of digitalization more closely to changes in communications infrastructure and the processing of social datafication can contribute to the theoretical debates on the issues surrounding the IoT. Questions that address the role of the physical and legal infrastructure of the IoT in the social order, as researchers who study the mediatization of politics do, should be at the center of government decisions. Especially in national plans and data protection policies, issues such as the legal and physical dependence that has been created need to be discussed more widely by society. Additionally, data storage services in 'clouds' impose a look at the datafication behind the tangled IoT´s infrastructure. A look at datafication implies observing the stages of growth and stabilization of these technologies. "The term not only captures a trend in the sense of changes that have already occurred, but it also manages to encapsulate expectations of its own stability and growth," (Hepp, 2019: 50).

Considering datafication by IoT technologies implies demystifying new agents that do not necessarily appear in the relationship between users and technology companies. Like the network of processors that handle users' data for a supposedly improved information service. In this comparative study, it was observed that the GDPR better details the procedures of controllers and data processors than the LGPD. Even in the face of the political instabilities that Brazil is facing in recent years, like the bill that attempts to postpone the beginning of LGPD until 2022, companies will have to adapt minimally to the new global requirements of security and privacy of information. Until then, the liability for users and subjects seems to prevail in terms of the use datafied by IoT. On August 26, 2021, the Brazilian Senate overturned an excerpt from the provisional bill that postponed the Law until 2021, meaning the third postponement in two years. According to the Folha de São Paulo newspaper, the text is to be sanctioned by President Jair Bolsonaro (without a party). The scenario points out that 38% of the companies interviewed by Ernst & Young, a consulting firm, in March 2020, claim to be in compliance with the Law. Even if the Law is sanctioned in 2020, the ANPD (National Data Protection Authority) will not be acting.

Slowly, virtual assistants are adhering with the LGPD/GDPR of complying with similar precepts of these laws in the southern hemisphere. However, attitudes on generalizing specific terms or not being explicit on how procedures are in accordance with legal aspects can be seen in the Terms of Use available to users. The advertisements that massify the need for a virtual assistant must be consistent with the urgent clarity of information on the subjects' data processing though. Alexa's products **try** to comply with the regulations imposed by LGPD, but some gaps in transparency about the data collected are still puzzling. In the meantime, national data protection policies in countries such as Brazil seem to be more state propaganda than rational business conduct.

Notes

1 The documentary analysis made in this research considered the English version of the GDPR.
2 Available at: <https://www.camara.leg.br/noticias/626827-PROPOSTA-ADIA-PARA-2022-A-VIGEN-CIA-DA-LEI-GERAL-DE-PROTECAO-DE-DADOS-PESSOAIS>. Accessed on February 28, 2019.

References

Amazon. 2019a. "Termos de Uso da Alexa." Last updated: October 30, 2019. Available at:< https://www.amazon.com.br/gp/help/customer/display.html?nodeId=201809740> Access on: 15 Feb. 19

Amazon. 2019b. "Condições de Uso." Last updated: October 30, 2019. Available at:< https://www.amazon.com.br/gp/help/customer/display.html?nodeId=201909000> Access on: 16 Feb. 19

Amazon. 2019c. "Perguntas Frequentes sobre a Alexa e os Dispositivos Alexa." Last updated: October 30, 2019. Available at:< https://www.amazon.com.br/gp/help/customer/display.html?nodeId=201602230> Access on: 16 Feb. 19

Averbeck-Lietz, Stefanie. 2015. "Eliseo Verón Leído Desde la Perspectiva de los Estudios en Comunicación Alemanes: Semio-Pragmática: Comunicación E Investigación En Mediatización." *Estudios - Centro de Estudios Avanzados. Universidad Nacional de Córdoba.* https://doi.org/10.31050/1852.1568.n33.11609.

Cavalcante, Fernando Luiz Nobre. 2019. "Vínculos de ancoragens e enquadramentos temáticos: Olhares itinerantes às interações midiatizadas em grupo". (Doctoral Degree). Programa de Pós-graduação em Estudos da Mídia (PPgEM), Universidade Federal do Rio Grande do Norte. https://repositorio.ufrn.br/jspui/hand-le/123456789/28635.

Cavalcante, Fernando Luiz Nobre, and Michael Manfred Hanke. 2020. "Framing Interaction Anchorage in Mediatized Groups." *Eikon* 1.7.

Chung, Hyunji, Michaela Iorga, Jeffrey Voas, and Sangjin Lee. 2017. "Alexa, Can i Trust You?" *Computer.* https://doi.org/10.1109/MC.2017.3571053.

Couldry, Nick, and Andreas Hepp. 2013. "Conceptualizing Mediatization: Contexts, Traditions, Arguments." *Communication Theory.* https://doi.org/10.1111/comt.12019.

Couldry, Nick, and Andreas Hepp. 2018. "The Continuing Lure of the Mediated Centre in Times of Deep Mediatization: Media Events and Its Enduring Legacy." *Media, Culture and Society.* https://doi.org/10.1177/0163443717726009.

van Dijck, J.. 2014. "Datafication, Dataism and Dataveillance." *Surveillance & Society.*

Flint, David. 2017. "Who's Listening to You?" *Business Law Review.*

Frazao, Ana. 2019. "Objetivos e Alcance da Lei Geral de Proteção de Dados." *Lei Geral De Proteção de Dados Pessoais e suas Repercussões no Direito Brasileiro.*

Furey, Eoghan, and Juanita Blue. 2018a. "Alexa, Emotions, Privacy and GDPR." https://doi.org/10.14236/ewic/hci2018.212.

Furey, Eoghan, and Juanita Blue. 2018b. "She Knows Too Much-Voice Command Devices and Privacy." In *29th Irish Signals and Systems Conference, ISSC 2018.* https://doi.org/10.1109/ISSC.2018.8585380.

General Data Protection Regulation (GDPR). 2018. *General Data Protection Regulation (GDPR)* – Final text neatly arranged. [online] Available at: https://gdpr-info.eu/ [Accessed Feb 18 2020].

Gregorio, Fernando, Gustavo González, Christian Schmidt, and Juan Cousseau. 2020. "Internet of Things." In *Signals and Communication Technology.* https://doi.org/10.1007/978-3-030-32437-7_9.

Grisot, Miria, Elena Parmiggiani, and Hanne Cecilie Geirbo. 2018. "Infrastructuring Internet of Things for Public Governance." In *26th European Conference on Information Systems: Beyond Digitization - Facets of Socio-Technical Change, ECIS 2018.*

Hasebrink, Uwe, and Andreas Hepp. 2017. "How to Research Cross-Media Practices? Investigating Media Repertoires and Media Ensembles." *Convergence.* https://doi.org/10.1177/1354856517700384.

Helberger, Natali, Jo Pierson, and Thomas Poell. 2018. "Governing Online Platforms: From Contested to Cooperative Responsibility." *Information Society.* https://doi.org/10.1080/01972243.2017.1391913.

Hepp, Andreas. 2011. "Medienkultur Als Die Kultur Mediatisierter Welten." In *Medienkultur.* https://doi.org/10.1007/978-3-531-94113-4_4.

Hepp, Andreas. 2019. *Deep Mediatization: Key Ideas in Media & Cultural Studies* (Kindle edition). Taylor & Francis.

Hepp, Andreas, Andreas Breiter, and Uwe Hasebrink. 2018. "Communicative Figurations: Transforming Communications in Times of Deep Mediatization." *Transforming Communications - Studies in Cross-Media Research*. https://doi.org/10.1007/978-3-319-65584-0.

Hintz, A., Dencik, L. & Wahl-Jorgensen, K. 2019. *Digital citizenship in a datafied society*. Polity Press, Cambridge.

Kelsen, Hans. 1966. "On The Pure Theory of Law." *Israel Law Review*. https://doi.org/10.1017/s0021223700013595.

Khan, Minhaj Ahmad, and Khaled Salah. 2018. "IoT Security: Review, Blockchain Solutions, and Open Challenges." *Future Generation Computer Systems*. https://doi.org/10.1016/j.future.2017.11.022.

Lee, Gyu Myoung, Noel Crespi, Jun Kyun Choi, and Matthieu Boussard. 2013. "Internet of Things." *Lecture Notes in Computer Science (Including Subseries Lecture Notes in Artificial Intelligence and Lecture Notes in Bioinformatics)*. https://doi.org/10.1007/978-3-642-41569-2-13.

Lei Geral de Proteção de Dados Pessoais Law n°. *13.709 (LGPD)*. 2018. *Lei Geral de Proteção de Dados Pessoais (LGPD)* – Final text neatly arranged. [online] Available at: https://www.planalto.gov.br/ccivil_03/_ato2015-2018/2018/lei/L13709compilado.htm [Accessed Feb 17 2020].

Mc Cullagh K., O. Tambou, S. Bourton (Eds.). February 2019. *National Adaptations of the GDPR, Collection Open Access Book*, Blogdroiteuropeen, Luxembourg, 130. Available at: https://wp.me/p6OBGR-3dP

Marco Civil da Internet. 2014. *Câmara dos Deputados, Brasil*, Edições Câmara, Brasilia.

Mendes, Laura Schertel, and Danilo Doneda. 2018. "Reflexões Iniciais Sobre a Nova Lei Geral de Proteção de Dados." *Revista de Direito Do Consumidor*.

Ni Loideain, Nora, and Rachel Adams. 2018. "From Alexa to Siri and the GDPR: The Gendering of Virtual Personal Assistants and the Role of EU Data Protection Law." *SSRN Electronic Journal*. https://doi.org/10.2139/ssrn.3281807.

Nobre Cavalcante, Fernando. 2017. "Vigilia y Vigilancia: Análisis de Contenido Del Registro Nacional de Acceso a Internet En Brasil." *Retos*. https://doi.org/10.17163/ret.n14.2017.03.

Powell, Alison. 2014. "Datafication, Transparency and Good Governance of the Data City." *Digital Enlightenment Yearbook 2014: Social Networks and Social Machines, Surveillance and Empowerment*.

Seo, Junwoo, Kyoungmin Kim, Mookyu Park, Moosung Park, and Kyungho Lee. 2017. "An Analysis of Economic Impact on IoT under GDPR." In *International Conference on Information and Communication Technology Convergence: ICT Convergence Technologies Leading the Fourth Industrial Revolution, ICTC 2017*. https://doi.org/10.1109/ICTC.2017.8190804.

Seo, Junwoo, Kyoungmin Kim, Mookyu Park, Moosung Park, and Kyungho Lee. 2018. "An Analysis of Economic Impact on IoT Industry under GDPR." *Mobile Information Systems*. https://doi.org/10.1155/2018/6792028.

Wachter, Sandra. 2018. "The GDPR and the Internet of Things: A Three-Step Transparency Model." *Law, Innovation and Technology*. https://doi.org/10.1080/17579961.2018.1527479.

Zanella, Andrea, Nicola Bui, Angelo Castellani, Lorenzo Vangelista, and Michele Zorzi. 2014. "Internet of Things for Smart Cities." *IEEE Internet of Things Journal*. https://doi.org/10.1109/JIOT.2014.2306328.

Websites

https://www.amazon.com/gp/help/customer/display.html?nodeId=201809740
https://www.amazon.com/gp/help/customer/display.html?nodeId=201909000
https://www.amazon.com/gp/help/customer/display.html?nodeId=201602230
https://www.camara.leg.br/noticias/626827-PROPOSTA-ADIA-PARA-2022-A-VIGENCIA-DA-LEI-GERAL-DE-PROTECAO-DE-DADOS-PESSOAIS

12

Improving the Security of Data in the Internet of Things by Performing Data Aggregation Using Neural Network-Based Autoencoders

Ab Rouf Khan, Mohammad Khalid Pandit, and Shoaib Amin Banday

CONTENTS

12.1 Introduction to the IoT and Smart City

IoT is a novel technological standard added to the ever-increasing technical arena of the present century. The essential components of any IoT framework include the sensor devices at the physical level of the three-tier architecture, which collect the data from the various applications and send it to the next level: network and application layers for further processing. Devices in the IoT are revamped from being 'traditional' to 'smart' by using the fundamental architectural components: sensor networks, embedded devices, ubiquitous and pervasive computing, communication technologies, internet protocols and applications (Li et al., 2015; Alaba et al., 2017). There are a considerable number of nodes in the IoT that gather the data from the IoT devices. These devices are usually autonomous. Amid the key characteristic features of IoT is heterogeneity and existence of distributive nature, which leads to several of the IoT applications being in requisite of statistics that are distributed on diverse nodes of data. The devices interchange the statistics and collaborate to finish the allotted responsibilities (Gubbi et al., 2013; Zhu et al., 2017).

The concept of the smart city, like the many other recent innovations in the technological arena, is quite new in nature. It is evolving alongside the IoT as the basic framework and conceptualization of the smart cities is hugely dependent on IoT. Although there is no concrete commonly recognized description of a smart city, the most appropriate definition can be:

> A city that monitors and integrates conditions of all of its critical infrastructures, including roads, bridges, tunnels, rails, subways, airports, seaports, communications, water, power, even major buildings, can better optimize its resources, plan its preventive maintenance activities, and monitor security aspects while maximizing services to its citizens.

(Hall et al., 2000; Chourabi et al., 2012)

The conceptualization and management of critical infrastructures are one of the most critical and challenging tasks to carry out in the smart city framework. The smart city mainly comprises of the elements: smart transportation, smart buildings, smart health, smart agriculture, smart parking, smart traffic management system, smart government, smart safety and surveillance systems, and smart grid utilities (Ramirez et al., 2016; Alavi et al., 2018). Figure 12.1 presents the various constituents of a smart city.

Aggregating data in the smart city architecture comprises collecting data from various constituents of the smart city architecture and expressing it in a summarized form. Implementing data aggregation mechanisms increases the efficiency of the overall scenario of collecting, transmitting and analyzing the data in the smart city architecture. Apart from the efficiency, the security is enhanced as well. Data aggregation approaches are employed at the physical level of the three-tier architecture of IoT to gather and aggregate the data from diverse source nodes from various smart city components in an effective fashion to improve the QoS parameters of the system comprising: network lifetime, traffic bottleneck, data accuracy, and energy consumption. Despite the heterogeneity of the various components of smart city architecture, the data in smart city applications (smart energy, smart health, smart agriculture, smart government etc.) could be pooled, amalgamated, matched and interrelated effortlessly to meet the requests and necessities of people. Smart cities being a combination of different components generating the data massive in size and structure at a very rapid rate have to possibly do much more with the efficiency of the data aggregation technique employed than any other IoT scenario (Pourghebleh and Navimipour, 2017; Zhang et al., 2018). Figure 12.2 demonstrates the common setup of data aggregation process in smart cities.

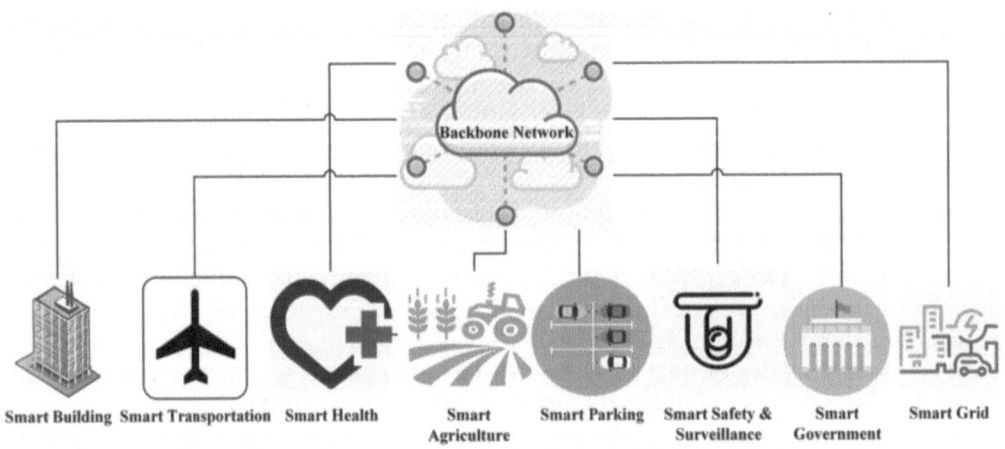

FIGURE 12.1
Constituents of a smart city.

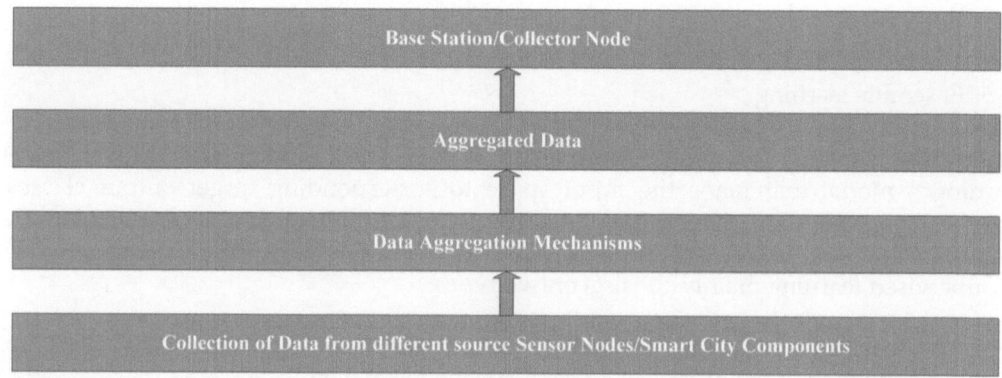

FIGURE 12.2
Broad picture of data aggregation in the IoT architecture.

12.2 Introduction to Neural Networks

Machine learning is a science of attaining the computers to act without being explicitly programmed. According to (Mitchel and Tom M., 1997) learning can be defined as "A program is said to learn from experience **E** concerning some task **T** and performance measure **P**, if its performance at task **T** as measured by **P**, improves with experience **E**." A learning algorithm is supposed to learn from data (i.e., data acts as an experience (**E**), and the task (**T**) is usually the classification or regression). The performance measure (**P**) is often measured as the accuracy measure achieved by the algorithm for the specific task at hand.

Traditional machine learning approaches are restricted to process the data in its raw form (i.e., it requires careful and deep domain capability to plan feature extractors that transfer the raw data into workable illustration or feature vectors). Modern machine learning techniques (using representational learning methodology) allow algorithms to be fed with raw inputs, and automatically convert them into feature vectors. Neural networks (deep learning) is a rep learning algorithm which has multiple levels of feature representation followed by a classifier. Deep learning encompasses repeated linear and nonlinear transformations of input data (more abstract features at every level). The success of deep learning algorithms lies in these transformations; enough of these transformations help to determine a very complex function. The fundamental feature of deep learning algorithms is that the various layers of features are not handcrafted (as in traditional ML) but are learned automatically from data through a general-purpose learning algorithm (gradient descent).

Deep learning is making significant progresses in various fields of science and engineering such as object detection (LeCun, Yann, et al., 1989; Krizhevsky et al., 2012; Zhao, Zhong-Qiu, et al., 2019; Liu, Li, et al., 2020), natural language processing (Manning et al., 2017; Young, Tom, et al., 2018), speech recognition (Noda, Kuniaki, et al., 2015; Yu, Dong, and Li Deng., 2016), medical diagnosis (Xu, Yan, et al., 2014; Ker, Justin, et al., 2017; Litjens, Geert, et al., 2017; Shen et al., 2017; Zhou, S. Kevin et al., 2017) and many more.

Machine learning algorithms are broadly classified into the following categories:

1. Supervised learning
2. Unsupervised learning

3. Reinforcement learning
4. Transfer learning
5. Ensemble learning

Supervised learning: It is the simplest yet most used learning methodology. It involves learning a model that maps the input space to corresponding target output classes. Applications, where the training data involves samples of the input vectors (X) alongside their matching target vectors (Y) are recognized as supervised learning problems.

Supervised learning mainly consists of two types:

a. **Classification:** problems where the output vector corresponds to class labels, i.e. prediction of the discrete class label for the input data.
b. **Regression:** problems where the output vector corresponds to continuous numeric values.

Unsupervised learning: in this type of learning only the input vector (X) is available without output or target labels. It involves finding similar traits in data and grouping them (e.g., Google news which finds news on related topic and groups together). The main types of unsupervised learning are:

a. **Clustering:** problem involves finding groups in data.
b. **Density estimation:** problem involves finding the distribution of data.

Reinforcement learning: learning algorithm where the agent learns to operate in the environment based on the feedback received. The goal of the RL algorithm is to make wiser decisions based on the feedback; the positive feedback corresponds to the good decision, and negative feedback represents the bad decision. In complex domains, RL is the only feasible way of training the program and achieve better results (e.g., AI-based game playing agents are mostly trained using RL algorithms).

Transfer learning: this type of learning involves the training of model on some baseline task, and then some or all of the trained model is used as the starting point of the next task at hand (e.g., we train the convolutional neural network on 1000 output class image-net dataset (P1) and then this pre-trained network can be used as a starting point for a different set of task P2). We can freeze the initial set of layers in the network and only train with P2 on few layers as P2 has fewer instances than P1. This method of learning has shown good generalization capability with minimum training time on P2.

Ensemble learning: Ensemble learning involves fitting the data on two or more models and combining the predictions from each model.

12.3 Machine Learning and IoT

In the year 2000, the average domestic monthly data usage was about 10 Gb, which increased ten times to around one petabytes in 2005. The usage increased to 3.7 exabytes in 2015, and it is estimated that in 2020 around 30 exabytes of data will be used monthly.

Such huge data will require a lot of energy for transmission from devices to cloud for computation. To put this in perspective, transferring 1 bit of data over the cellular network requires around 500 (μj) micro joules of energy. For 30 exabytes (EB) data it will require:

$$30\,EB \times 500\left(\mu j\right) = 500\,terra\,watt - hours. \tag{12.1}$$

This is roughly equivalent to 2% of the world's electricity consumption.

The similar trend is being followed in the IoT framework where the number of ever-increasing devices leads to the generation of humongous data, which if sent to the base station for making the inferences without being processed at the physical and network layer is going to consume a lot more bandwidth, and the other resources like the battery life of the sensors. Thus, the processing of data, which mainly can be performed by aggregating the data at the physical layer can drastically improve the QoS factors of the IoT network. To perform data aggregation, various strategies can be implemented, and neural network-based autoencoders can be an efficient way to deal with the challenge of aggregating the data efficiently. Thus, the usage of neural networks in the IoT is of utmost importance, and we are exploiting the same in this chapter.

12.4 Need for Dimensionality Reduction

For the enormous data generated by IoT devices, the energy requirements for transferring the aggregate data to processing engines housed in fog or cloud (MK Pandit, et al., 2017) is very costly. Equation 12.1 indicates the data transfer cost, which is likely to be utilized for transferring such huge volumes of data.

To extract useful information (decision) from the data, the features are derived from the input data and then fed into the computational algorithm. Usually, these features are correlated, hence redundant, and therefore could be omitted as they won't be useful in decision making. It seems like the glimmer of hope of the IoT applications whereby not all data generated by these devices is to be sent to the processing engines. This is where dimensionality reduction comes into action. It is the method of decreasing the dimensions of the input data such that only the relevant information is transferred which is usually less than the actual dimension of the input data. Figure 12.3 represents the dimensionality reduction.

In this chapter, we will discuss the use of neural network-based techniques for dimensionality reduction. A particular type of neural network algorithm called autoencoders can be used effectively in the IoT domain which will serve the purpose of dimensionality reduction at the data source, i.e. IoT device itself. A part of autoencoder called 'ENCODER' will be used at the IoT device which will encode the data in lower dimension and then send it over the communication network to the processing engine hosted in the cloud or fog, thereby reducing the load on the communication infrastructure as well as the cost associated with the communication. At the other end (fog and cloud) the other part of the autoencoder 'DECODER' will reconstruct the input received from the IoT device (encoder). The decoder will approximate the actual information, thereby making the decision-making process as accurate as possible.

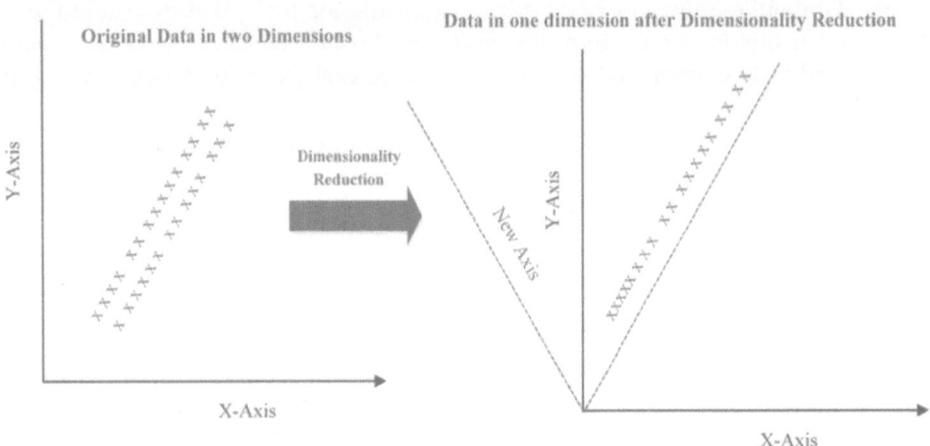

FIGURE 12.3
Dimensionality reduction.

12.5 Autoencoders

An autoencoder is a neural network that is intended to set the target values (Y) to be equal to input (X) (i.e., if X is the input to the autoencoder and Y represents the output produced). Let $A(X)$ denote the autoencoder function, therefore, autoencoder tries to learn the function $A_{w,b}(X) = X$, where w, b are the parameters of the model. The function $A(.)$ seems to be trivial to learn unless certain constraints are applied to it, which makes it particularly useful. One such restriction is to limit the number of hidden layers or number of nodes in hidden layers, e.g. let X be an image of size 20×20 given as input to the autoencoder, therefore $X \in R^{400}$. Since the output of the network also must be same as input (i.e. $Y \in R^{400}$), if we will force the hidden layer to have only 100 units, therefore, we can learn the compressed representation of the input at the hidden layer. Also, we can reconstruct the input at the output layer. Internally the autoencoder has hidden layer representation (code h) that is the representation of the input. In the above example, the input was R^{400} and the hidden representation (h) was R^{100}. From the above example, we can determine the autoencoders consists of two parts (i.e. encoder and decoder). The encoder function $h = f(\overline{X})$ produces the code (i.e., hidden representation and the decoder function) $Y = g(h)$ produces the reconstruction of the input. The goal of the autoencoder is to learn the function $g(f(X)) = X$. Figure 12.4 represents the autoencoders.

12.6 Dimensionality Reduction with Autoencoders

The idea of autoencoders has been there in machine learning literature for decades (Goodfellow, Ian et al., 2016). In the IoT scenario, autoencoders could be widely used for dimensionality reduction and feature engineering. In the context of dimensionality

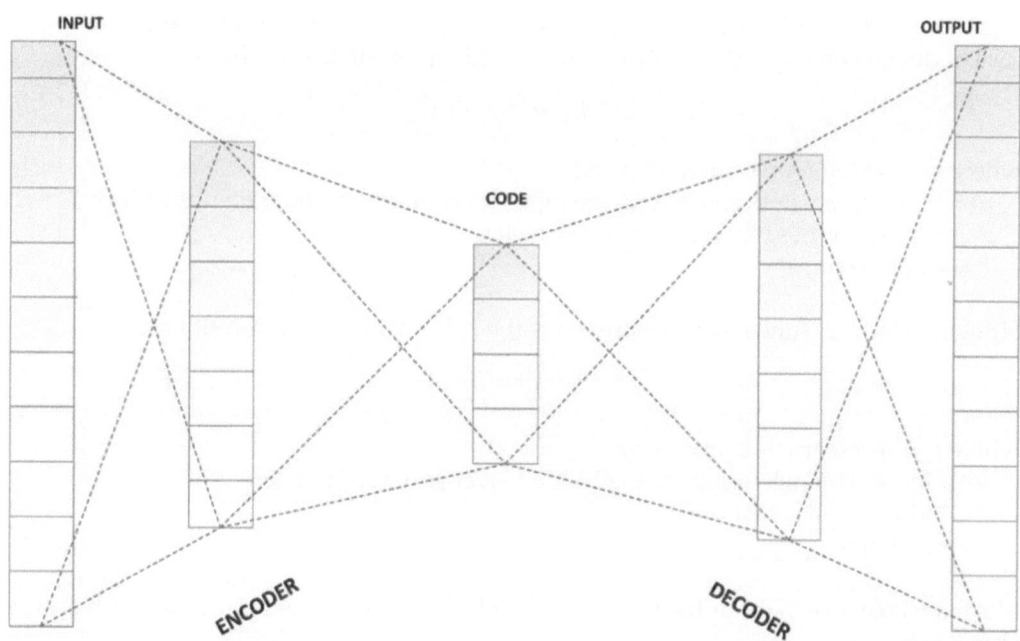

FIGURE 12.4
Autoencoders.

reduction, autoencoders can be thought of as compression and decompression algorithm, wherein the IoT data produced can be compressed at the source aggregated and then forwarded to the processing engines housed in the cloud. The compressed data can be decompressed at the cloud. It must be noted that the decompression is lossy. The compression and decompression functions are implemented via neural networks and have the following properties.

1. **Data specific:** autoencoders are data specific (i.e., they will be successfully able only to compress and decompress the similar data on which they are trained).
2. **Lossy compression de-compression:** autoencoders produce the lossy reconstruction of the input.
3. **Learned automatically (trained on a specific dataset):** autoencoders learn automatically from the data (i.e. they are trained using some learning algorithm which helps to learn the parameters of the model by gradient-based algorithm; usually gradient descent or its variants).

The compression function is the encoder part of the model, and the decompression represents the decoder. To build an autoencoder, three things are of paramount importance:

1. Encoder function.
2. Decoder function.
3. Loss function.

Encoder function $g(.)$ converts the input sample x_i into hidden representation (h) (i.e., in case of dimensionality reduction the compressed representation of input).

$$h = g\left(wx_i + b\right)$$

where h = hidden representation (code),
  w = set of parameters connecting the input layer of autoencoder to hidden layer,
  x_i = current input sample fed to autoencoder,
  b = set of bias

Similarly, decoder function $f(.)$ reconstructs the input again at the output layer, thus

$$\overline{x}_i = f\left(w^* h + c\right)$$

where $\overline{x}_i$ = reconstructed form of x_i,
  w^* = set of parameters connecting hidden layer and output layer,
  h = hidden representation,
  c = set of bias

The main objective of the autoencoder is to minimize a specific loss function which ensures that $\overline{x}$ to x_i.

12.6.1 Choice of Encoder and Decoder Functions

Case 1: the case in which the input to autoencoder is of binary form (i.e. $x_i = \{0,1\}$ for the decoder function $f(.)$), we use the logistic function as it reconstructs the output between 0 and 1 (i.e. $\overline{x}_i = logistic\left(w^* h + c\right)$ for such cases the encoder function is typically chosen to be sigmoid).
  Case 2: when the input to the autoencoder is continuous (real-valued), (i.e., $x_i \in R$ for the decoder function $f(.)$), we choose the linear function, i.e. $\overline{x}_i = \left(w^* h + c\right)$. Yet again the encoder function can be selected to be any function, but typically sigmoid function is used.

12.6.2 Choice of Loss Functions

Case 1: In case the input to the encoder is binary, $x_i = \{0,1\}$. Here we have decoder function $f(.)$ as logistic function (i.e. the decoder will produce the outputs between 0 and 1 and thus can be interpreted as probabilities). Therefore, the loss function used here is cross-entropy loss, and is determined as:

$$min\left\{ -\sum_{j=1}^{n} x_{ij} \log \overline{x}_{ij} + \left(1 - x_{ij}\right) \log\left(1 - \overline{x}_{ij}\right) \right\}$$

The function will attain the minimum value when $x_{ij} = \overline{x}_{ij}$.
  Case 2: in case the input to the autoencoder is continuous (real-valued) (i.e. $x_i \in R$, here the decoder function is a linear function). Squared error loss is shown to work best for such cases. Therefore, the loss function will simply be:

$$min_{w,w^*} \frac{1}{m} \sum_{i=1}^{m} \sum_{j=1}^{n} \left(\overline{x}_{ij} - x_{ij}\right)^2$$

$$min_{w,w^*} \frac{1}{m} \sum_{i=1}^{m} \left(\overline{x}_i - x_i\right)^T \left(\overline{x}_i - x_i\right)$$

This is the vector representation of the squared error loss function.

12.7 Experimental Results

We performed experiments on two benchmark datasets MNIST (Y. LeCun et al., 1998) and labeled faces in the wild LFW (Erik Learned-Miller et al., 2007). MNIST is a database of 70,000 greyscale images of handwritten digits. The dimensions of MNIST images is 28×28 pixels. LFW dataset is a benchmark for face recognition. Over 13,000 images collected from the web constitute this data set. Experiments are conducted on various types of autoencoder models; viz. FFNN-based autoencoder, convolutional neural network-based autoencoder, denoising autoencoder etc. All the experiments are conducted using TensorFlow (Abadi, Martín, et al., 2016).

1. MNIST
    a. FFNN-based autoencoder: this is the simplest form of the autoencoder. Here we created a 3-layer FFNN with 784 units in both input and output layers. In this experiment, we have reduced the dimensionality in the hidden layer to 32 dimensions. Here we use the ReLU in the encoder and sigmoid in the decoder. The network is trained for 50 epochs with the mini-batch size of 256. Figure 12.5 shows the original input and the reconstructed input produces by this network. The training and validation loss achieved are *0.106* and *0.104*, respectively.
    b. CNN based autoencoder: in this model, we have three convolutional layers in both encoder as well as a decoder. Max-pooling layers follow all the convolution layers. The activations function used in convolutional layers is ReLU and the decoder function used is sigmoid. We use 16 3×3 filters in input as well as output layers. All the hidden convolutional layers apply 8 3×3 filters. Here the hidden representation size is reduced to 128 dimensions. The network is trained using Adam optimizer with cross-entropy as loss function. The network is trained for 50 epochs with a mini-batch size of 120. The training and validation loss obtained is *0.912* and *0.8667*, respectively (Figure 12.6).

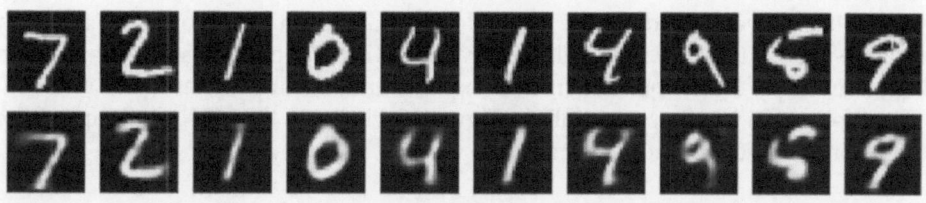

FIGURE 12.5
Row 1 represents the original input, and row 2 represents the reconstructed input (Feed Forward autoencoder).

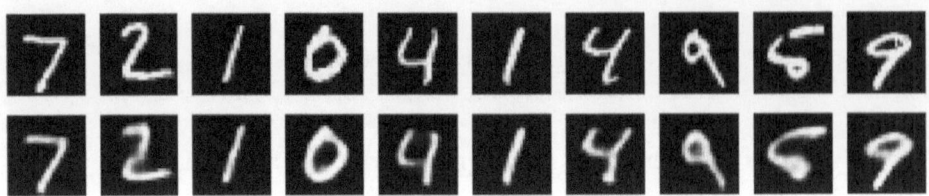

FIGURE 12.6
Row 1 represents the original input, and row 2 represents the reconstructed input (convolutional autoencoder).

2. LFW Dataset: On LFW dataset, we used a convolutional neural network-based auto-encoder with three layers in encoder as well as a decoder. Activation function used in all the layers is ReLU and the decoder function used is cross-entropy. In this model, we checked the impact of code size with the mean square error of input image and reconstructed image. We checked with various code sizes (i.e. 32-, 64-, 128- and 256-dimensional hidden representations) (Figures 12.7 and 12.8).

Figure 12.9 represents the training and validation loss on various code sizes. It represents the original input with hidden representation followed by a reconstructed input.

From Figure 12.9, it is clear that, as the code size increases, the reconstructed input tends toward the original. It represents the mean square error computed on five input images with reconstructed inputs. There is a trade-off between the compression size and the reconstructed error. Figure 12.9 shows that code dimensions 32 has the highest mean square error (MSE), and the MSE decreases as the code size increases. The code size of 256 has the minimum MSE and produces the best-reconstructed input.

12.7.1 Denoising Autoencoders

In this process, we corrupt the input using the probabilistic method. The corrupted input is fed into the model, and the model is expected to output the original un-corrupted input. Let x_i represent the original input, and $\tilde{x}_i$ is the corrupted form of input which is given as input to the autoencoder model. Here the loss function of the model will minimize the error between reconstruction $\bar{x}_i$ and original input x_i rather than $\tilde{x}$. It is particularly useful in IoT scenario where the data collection can be erroneous, and thus, the data is collected with some noise. This model can filter out the noise, thus reconstructing the noise-free data. Here the models are more robust to data as it cannot just make a travail mapping from input to output (reconstruction) (i.e., it captures the characteristics of data more correctly). Figure 12.10 represents the input given to the model of varying noise levels (increasing from left to right).

Here also we used a convolutional neural network-based autoencoder with three layers in encoder as well as a decoder. We added the Gaussian noise to training and validation data with $\sigma = 0.1$. The network was trained with 50 epochs using Adam optimizer and cross-entropy loss function. In this experiment, the input has been reduced to a 100-dimensional hidden representation. Figure 12.11 shows the corrupted input followed by the reconstructed output by the model.

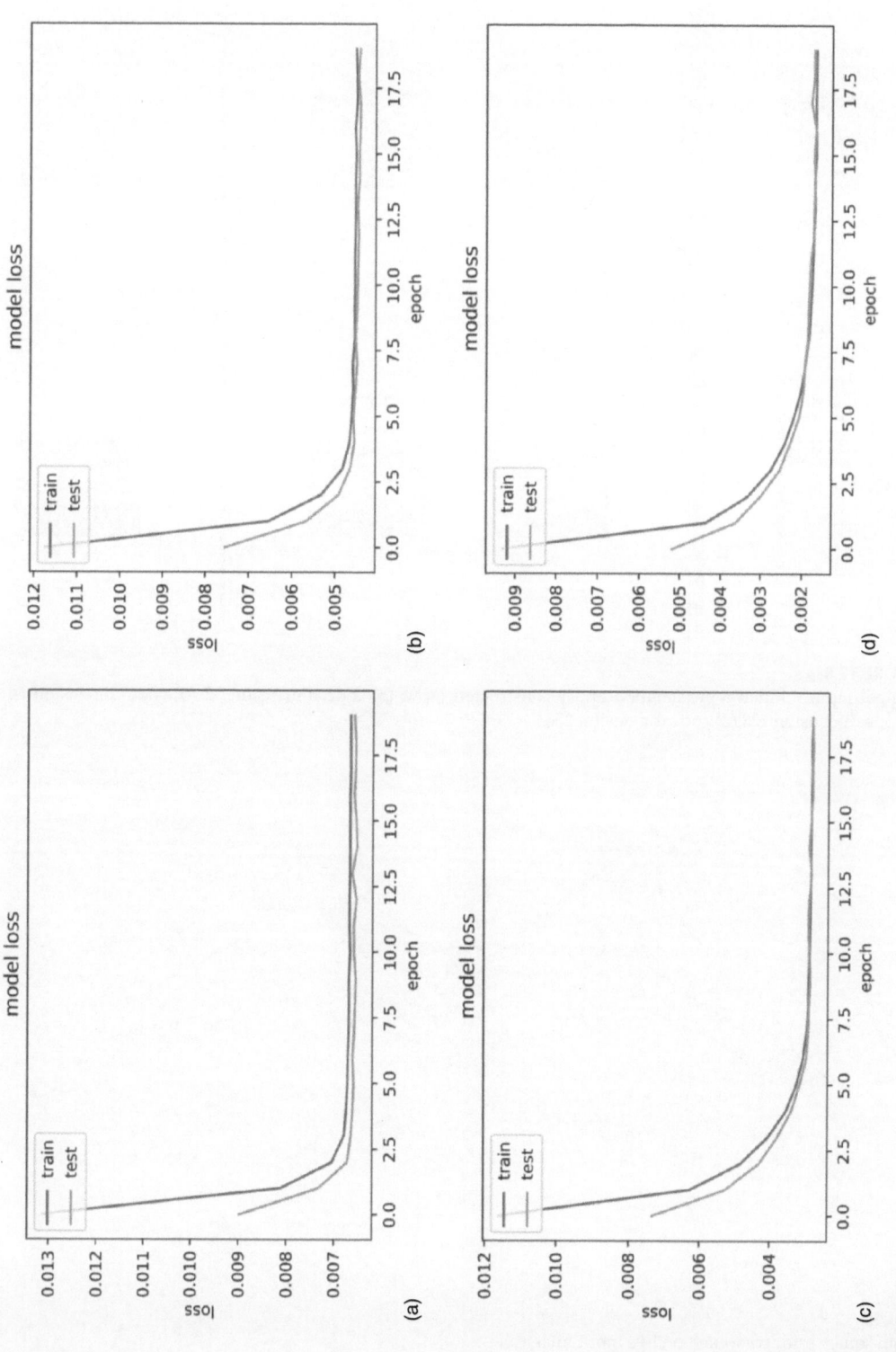

FIGURE 12.7
Validation and Training loss (a) code dimension 32 (b) code dimension 64 (c) code dimension 128 (d) code dimension 256.

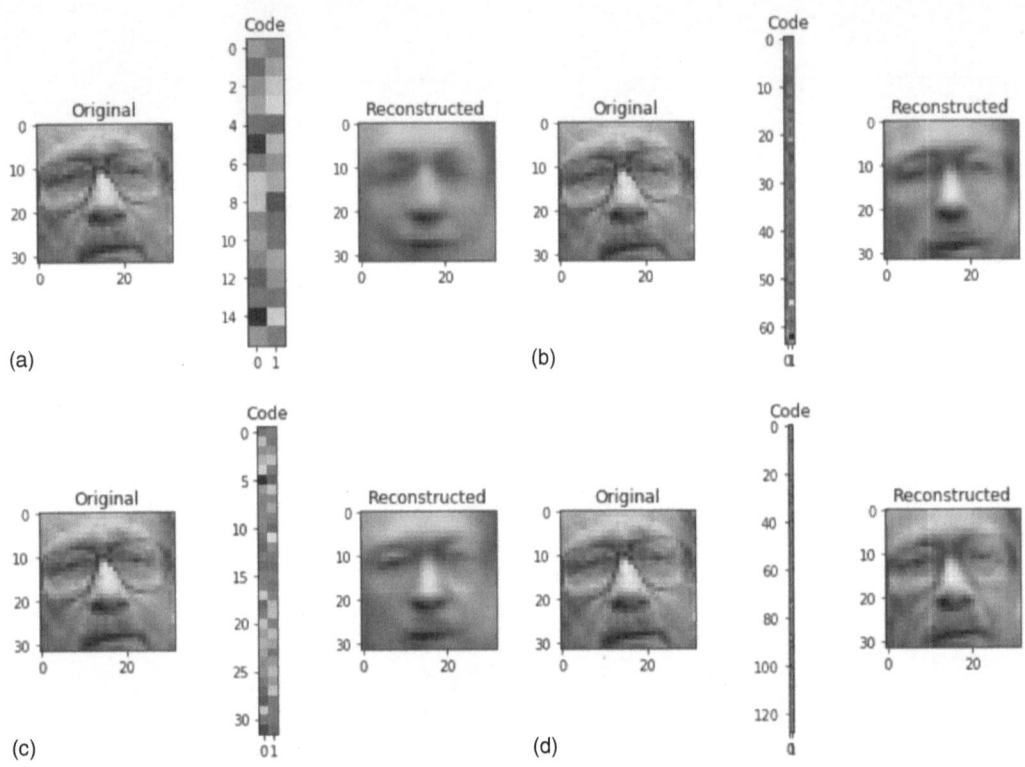

FIGURE 12.8
Original input, hidden representation and reconstructed input (a) code dimension 32 (b) code dimension 64 (c) code dimension 128 (d) code dimension 256.

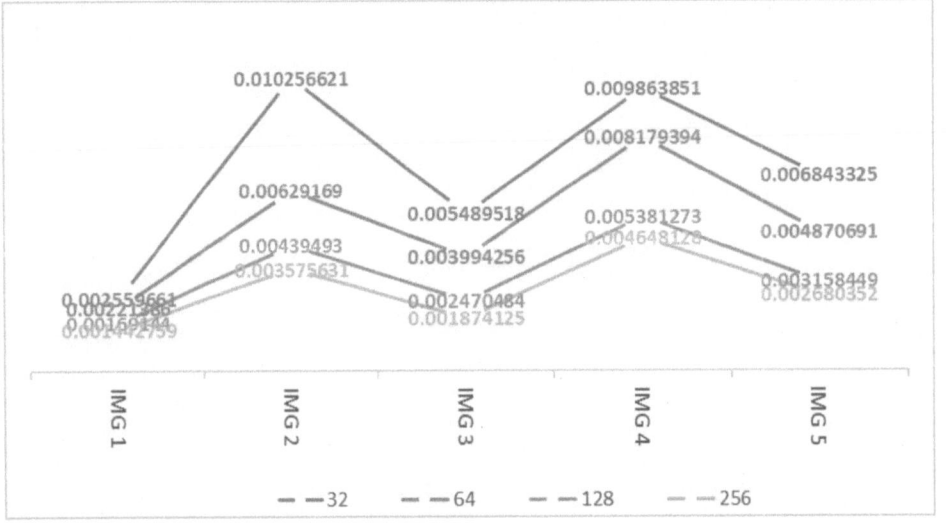

FIGURE 12.9
Mean square error computed on five input images.

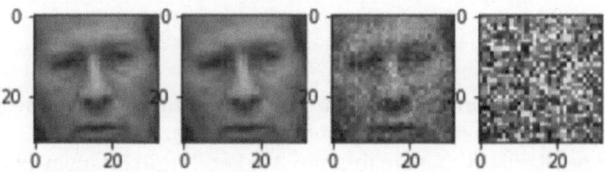

FIGURE 12.10
Corrupted input given to autoencoder. (Left to right) input with zero noise, subsequent inputs with added noise.

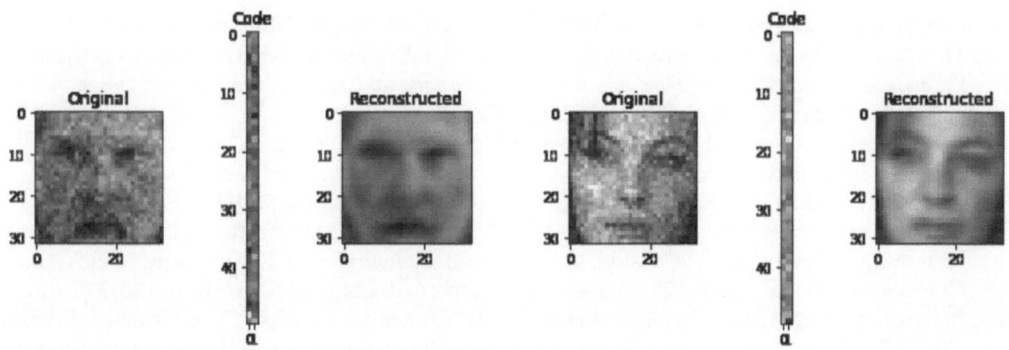

FIGURE 12.11
Output generated by denoising autoencoder.

12.8 Conclusion

The IoT can be beheld as a paradigm where the Internet-capacitate devices (actuators, sensors etc.) interconnect with each other to turn the physical world things into the digital world things. Data aggregation aims at collecting the data from various source nodes and sending only the summarized data. Aggregating the data not only increases the efficiency of the network on a whole but also improves security as well. We have used neural networks based autoencoders to perform the data aggregation. FFNN-based autoencoder, CNN based autoencoder and denoising autoencoder techniques have been implemented on MNIST and LFW datasets on code dimensions 32, 64, 128 and 256. Code dimension 32 has the highest MSE, and the MSE decreases as the code size increases. The code size of 256 has the minimum MSE and produces the best-reconstructed input.

Acknowledgment

This chapter was funded by TEQIP, IUST GRANT NO: IUST/TEQIP/19/36/53.

References

Abadi, Martín, et al. 2016. "Tensorflow: A system for large-scale machine learning." *12th {USENIX} Symposium on Operating Systems Design and Implementation ({OSDI} 16).*

Alaba, Fadele Ayotunde, Mazliza Othman, Ibrahim Abaker, Targio Hashem, and Faiz Alotaibi. April 2017. "Internet of Things Security: A Survey." *Journal of Network and Computer Applications* 88: 10–28. https://doi.org/10.1016/j.jnca.2017.04.002.

Alavi, Amir H., Pengcheng Jiao, William G. Buttlar, and Nizar Lajnef. July 2018. "Internet of Things-Enabled Smart Cities: State-of-the-Art and Future Trends." *Measurement: Journal of the International Measurement Confederation* 129: 589–606. https://doi.org/10.1016/j.measurement. 2018.07.067.

Chourabi, Hafedh, Taewoo Nam, Shawn Walker, J. Ramon Gil-Garcia, Sehl Mellouli, Karine Nahon, Theresa A. Pardo, and Hans Jochen Scholl. 2012. "Understanding Smart Cities: An Integrative Framework." *Proceedings of the Annual Hawaii International Conference on System Sciences*, 2289–2297. https://doi.org/10.1109/HICSS.2012.615.

Erik Learned-Miller, Gary B. Huang, Aruni RoyChowdhury, Haoxiang Li, and Gang Hua. 2007, "Labeled Faces in the Wild: A Survey." *Advances in Face Detection and Facial Image Analysis*

Goodfellow, Ian, Yoshua Bengio, and Aaron Courville 2016. *Deep Learning.* MIT Press.

Gubbi, Jayavardhana, Rajkumar Buyya, Slaven Marusic, and Marimuthu Palaniswami. 2013. "Internet of Things (IoT): A Vision, Architectural Elements, and Future Directions." *Future Generation Computer Systems* 29 (7): 1645–1660. https://doi.org/10.1016/j.future.2013.01.010.

Hall, Robert E, J Braverman, J Taylor, and H Todosow. September 28, 2000. "The Vision of A Smart City. In *2nd International Presented at the Life Extension Technology Workshop* Paris, France. https://doi.org/10.1017/CBO9781107415324.004.

Ker, Justin, et al. 2017. "Deep learning applications in medical image analysis." *IEEE Access* 6: 9375–9389.

Krizhevsky, Alex, Ilya Sutskever, and Geoffrey E. Hinton. 2012. "Imagenet classification with deep convolutional neural networks." *Advances in Neural Information Processing Systems.*

LeCun, Yann et al. 1989. "Backpropagation applied to handwritten zip code recognition." *Neural Computation* 1 (4): 541–551.

Li, Shancang, Li Da Xu, and Shanshan Zhao. 2015. "The Internet of Things : A Survey" 17 (April 2014): 243–259. https://doi.org/10.1007/s10796-014-9492-7.

Litjens, Geert et al. 2017. "A survey on deep learning in medical image analysis." *Medical Image Analysis* 42: 60–88.

Liu, Li et al. 2020. "Deep learning for generic object detection: A survey." *International Journal of Computer Vision* 128 (2): 261–318

Manning, Christopher, and Richard Socher. 2017. "Natural language processing with deep learning." *Lecture Notes Stanford University School of Engineering.*

Mitchell, Tom M. 1997. "Machine Learning." "Tata McGraw Hill Education".

Pandit MK, Mohammad Ahsan Chishti, Roohie Naaz Mir. November 2017. "Machine learning at the edge of IoT." *CSI Communications.*

Noda, Kuniaki et al. 2015. "Audio-visual speech recognition using deep learning." *Applied Intelligence* 42 (4): 722–737.

Pourghebleh, Behrouz, and Nima Jafari Navimipour. July 2017. "Data aggregation mechanisms in the Internet of Things: A systematic review of the literature and recommendations for future research." *Journal of Network and Computer Applications* 97: 23–34. https://doi.org/10.1016/j. jnca.2017.08.006.

Ramirez, Alejandro R. Garcia, Israel González-Carrasco, Gustavo Henrique Jasper, Amarilys Lima Lopez, Jose Luis Lopez-Cuadrado, Angel García-Crespo, Stefano Bresciani, et al. 2016. "Smart City Architecture and Its Applications Based on IoT." *Sensors (Switzerland)* 16 (11): 611–616. https://doi.org/10.1109/NGMAST.2016.17.

Shen, Dinggang, Guorong Wu, and Heung-Il Suk. 2017. "Deep learning in medical image analysis." *Annual Review of Biomedical Engineering* 19: 221–248.

Xu, Yan, et al. 2014. "Deep learning of feature representation with multiple instance learning for medical image analysis." *IEEE International Conference on Acoustics, Speech and Signal Processing (ICASSP)*. IEEE.

Y. LeCun, L. Bottou, Y. Bengio, and P. Haffner. November 1998. "Gradient-based learning applied to document recognition." *Proceedings of the IEEE*, 86 (11): 2278–2324

Young, Tom et al. 2018. "Recent trends in deep learning based natural language processing." *IEEE Computational Intelligence Magazine* 13 (3): 55–75.

Yu, Dong, and Li Deng. 2016. *Automatic Speech Recognition*. Springer London Limited.

Zhang, Ping, Jianxin Wang, Kehua Guo, Fan Wu, and Geyong Min. 2018. "Multi-Functional Secure Data Aggregation Schemes for WSNs." *Ad Hoc Networks* 69: 86–99. https://doi.org/10.1016/j.adhoc.2017.11.004.

Zhao, Zhong-Qiu et al. 2019. "Object detection with deep learning: A Review." *IEEE Transactions on Neural Networks and Learning Systems* 30 (11): 3212–3232.

Zhou, S. Kevin, Hayit Greenspan, and Dinggang Shen, eds 2017. *"Deep learning for Medical Image Analysis."* Academic Press.

Zhu, Tao, Sahraoui Dhelim, Zhihao Zhou, Shunkun Yang, and Huansheng Ning. 2017. "An architecture for aggregating information from distributed data nodes for Industrial Internet of Things." *Computers and Electrical Engineering* 58: 337–349. https://doi.org/10.1016/j.compeleceng.2016.08.018.

Appendices

Appendix A: Wireless Communication Channel Modeling

In this section, we briefly discuss the mathematical modeling of a wireless channel.

The behavior of the wireless medium toward electromagnetic radiation between transmitter and receiver with respect to time and frequency defines the channel in a wireless communication system. This behavior is broadly reflected in the following two effects:

- **Large-scale fading;** This effect usually occurs when the receiving nodes are mobile, due to which there are variations in signal with respect to distance and shadowing by large structures like buildings.

- **Small-scale fading;** This effect occurs due to the constructive and destructive interference of the multiple signal paths between the transmitter and receiver.

These channel effects can be modeled as a linear time-varying system with the transmitted signal $\varphi(t)$ as input and received signal $\sum_i a_i(f,t)\varphi\{t-\tau_i(f,t)\}$ as output. Where $a_i(f,t)$ and $\tau_i(f,t)$ are respectively the overall attenuation and propagation delay at time t from the transmitter to the receiver over the ith path. If it is assumed that $a_i(f,t)$ and $\tau_i(f,t)$ are frequency independent, then by superposition we can generalize the above input-output relation to an arbitrary input $x(t)$ as:

$$y(t) = \sum_i a_i(t) x\{t-\tau_i(t)\}$$

Since the channel is assumed to be linear, we can describe the input-output relationship in terms of its impulse response $h(\tau,t)$ as:

$$y(t) = \int_{-\infty}^{+\infty} h(\tau,t) x\{t-\tau\} d\tau$$

Where $h(\tau,t) = \sum_i a_i(t)\delta\{\tau-\tau_i(t)\}$ is the impulse response of the channel.

In wireless applications, the communication typically occurs in a pass-band $f_c - \dfrac{W}{2}$ $f_c - \dfrac{W}{2}$ with bandwidth W around a center frequency f_c. However, most of the signal processing operations are actually performed at the baseband. So from design point of view baseband equivalent representation of the system is more useful.

For any complex baseband signal $r_b(t)$ the real pass-band signal is given by $r(t) = Real[r_b(t)e^{j2\pi f_c t}]$. Thus if $x_b(t)$ and $y_b(t)$ are the complex baseband equivalents of the transmitted signal and received signal respectively, the baseband equivalent of the channel model are calculated as:

$$Real\left[y_b(t)e^{j2\pi f_c t}\right] = \sum_i a_i(t)Real\left[x_b\{t - \tau_i(t)\}e^{j2\pi f_c(t - \tau_i(t))}\right]$$

$$= Real\left[\left\{\sum_i a_i(t)x_b\{t - \tau_i(t)\}e^{-j2\pi f_c \tau_i(t)}\right\}e^{j2\pi f_c t}\right]$$

This gives:

$$y_b(t) = \sum_i a_i^b(t)x_b\{t - \tau_i(t)\}$$

Where; $a_i^b(t) = a_i(t)e^{-j2\pi f_c \tau_i(t)}$.

The corresponding discrete time model of the channel is given by:

$$y[n] = \sum_k h_k[n]x[n-k]$$

Where $h_k[n]$ is the kth complex channel filter tap at time n.

Since the wireless channel is often a noisy one, we include an additive white Gaussian noise process with zero mean and power spectral density $\frac{N_0}{2}$ into the channel model, so that complete model is described as:

$$y[n] = \sum_k h_k[n]x[n-k] + w(n)$$

Appendix B: Wiretap Coding for Physical Layer Security

In this section, we will first briefly re-visit the channel coding problem at physical layer. We recall the basic result of channel coding for point to point channel, i.e., a single transmitter and a single receiver communicating over a noisy channel. This general scenario is applicable for IoT nodes also, where both transmitter and receiver can be IoT node or one of them can be any other communication device and other one IoT node. C. E. Shannon in 1948 published his seminal paper and addressed this problem (Shannon, C.E., 1948). To solve the problem of transmission of information over noisy channel, Shannon modeled communication sources as random processes and channel was characterized in terms of conditional probability matrix, i.e., the output of channel was related to input symbol transmitted via a conditional probability. Shannon laid the foundation of new mathematical science called Information Theory to characterize the problem of communication in terms of probability theory. We first define some basic terms of information theory, which will be needed to understand the Wyner's wiretap coding.

Entropy: For an independently and identically distributed (iid) random sequence X_1, $X_2,...X_n$, with probability mass function $Pr(X_i = x_i) = p(x_i)$, $i = 1,...n$, Entropy for this process is defined as

$$H(X) = -\sum_{i=1}^{n} p(x_i) \log_2 p(x_i)$$

Remarks: Entropy is also defined as measure of randomness a source of information possesses. The operational interpretation of entropy is also given as the minimum number of bits required to represent a random source.

Conditional Entropy: Let X_i, $i = 1,...,n$ denote the sequence of random variables which denotes the input to the communication channel, Y_i be the received sequence corresponding to the channel input. Let $p(y_i | x_i) = Pr(Y_i = y_i | X_i = x_i)$ denote the conditional probability which characterizes the noisy channel. Then conditional entropy is defined as

$$H(Y|X) = -\sum_{i=1}^{n}\sum_{j=1}^{n} p(x_i, y_j) \log_2 p(y_j | x_i)$$

Mutual Information: Mutual information is defined as (Shannon, C.E., 1948), (Cover, T.M. and Thomas, J.A., 2012)

$$I(X;Y) = H(Y) - H(Y|X) = H(X) - H(X|Y) = \sum_{i=1}^{n}\sum_{j=1}^{n} p(x_i, y_j) \log_2 \frac{p(x_i, y_j)}{p(x_i), P(y_j)}$$

Now we state the Shannon's noisy channel coding theorem:

The maximum rate achievable such that the probability of error is driven to zero as block-length n tends to infinity is given by

$$C = \max_{p(x)} I(X;Y)$$

Wyners Wiretap coding: Aaron D Wyner in 1975 proposed an information theoretic coding scheme for single transmitter, single receiver and an eavesdropper channel model (also called **wiretap channel**) [22]. Using this coding/decoding scheme, one can achieve security at physical layer itself. In this scheme, the security of quantified by a quantity called equivocation, which is defined as

Equivocation: Let W be message which is chosen from a set $W = \{1,2,...,2^{nR}\}$ with uniform probability, and $Z^n = (Z_1,...,Z_n)$ be the codeword sequence received by an eavesdropper, then the uncertainty about the message W after observing Z^n is given by $H(W | Z^n)$, this quantity is called equivocation.

Another similar quantity which is used as measure of security is called leakage rate and is defined as:

Leakage Rate: Leakage rate is defined as

$$R_L = \frac{1}{n} I(W;Z^n)$$

Perfect Secrecy: In information theoretic security setup, we say that a message W is perfectly secure if $R_L \rightarrow 0$ as $n \rightarrow \infty$ or equivalently $H(W \mid Z^n) \rightarrow H(W)$.

Following is the main result from **Wyner's wiretap coding**:

The achievable rate while satisfying perfect secrecy constraint at eavesdropper and driving probability of error to zero at legitimate receiver is given by

$$R < C = \max_{p(x)} \left[I(X;Y) - I(X;Z) \right] = C_1 - C_2$$

Where $C_1 = \max_{p(x)} I(X;Y)$ and $C_2 = \max_{p(x)} I(X;Z)$

Sketch of proof: Instead of assigning codewords to messages, 2^{nC_1} codewords are divided into sub-codebooks each containing 2^{nC_2} codewords,

Hence a total of $2^{nC_1}/2^{nC_2}$ sub-codebooks are generated.

The number of messages are chosen to be $2^{n(C_1 - C_2)}$.

To transmit any message, a codeword is chosen at random from a sub-codebook corresponding to this message.

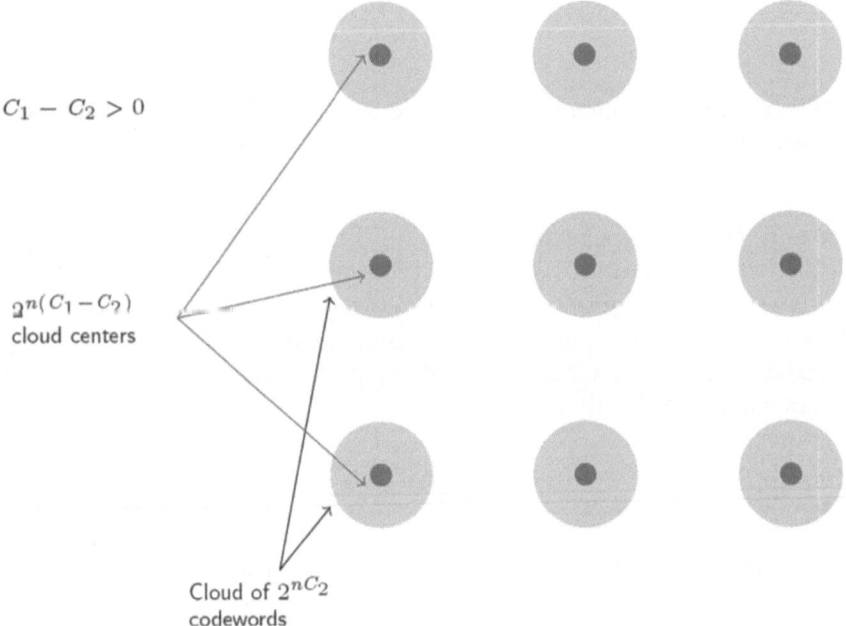

$C_1 - C_2 > 0$

$2^{n(C_1 - C_2)}$
cloud centers

Cloud of 2^{nC_2}
codewords

Eavesdropper can decode the codeword within sub-codebook,

Since the rate for this decoding is C_2, but according to Shannon's Channel coding theorem, eve cannot decode from which sub-codebook the codeword was chosen (hence cannot know which message was transmitted). The rate achievable is

$$\frac{\log_2 2^{n(C_1 - C_2)}}{n} = C_1 - C_2$$

Corollary: For Additive White Gaussian Noise (AWGN) wiretap channel, modeled as

$$Y = X + N_1, \quad Z = X + N_2$$

Where X is the input codeword to the channel, N_1 is the Gaussian noise with mean 0 and variance σ_1^2 and N_2 is Gaussian noise added at eavesdropper's receiver with mean 0 and variance σ_2^2. Now we recall the result from [24] that the maximum secrecy rate achievable for AWGN wiretap channel is given by

$$C_s = \frac{1}{2}\left[\log_2\left(1 + P/\sigma_1^2\right) - \log_2\left(1 + P/\sigma_2^2\right)\right]$$

Appendix C: Constrained Optimization and KKT Conditions

For general non-linear programming problems with equality and inequality constraints, KKT conditions are basically the first derivative tests that form the necessary conditions for a solution of a problem to be optimal. As in Lagrange multiplier method, the constrained optimization problem is written as a Lagrange function and the stationary points are obtained by the first derivative tests. Consider the following non-linear programming problem

$$\min f(x)\, s.t : g(x) \le 0 \, and \, h(x) = 0$$

Here, $f(x)$, $g(x)$ *and* $h(x)$ are continuously differentiable functions. The Lagrangian function for the above defined constrained problem is written as:

$$L(x,\lambda,v) = f(x) + \lambda g(x) + vh(x)$$

Now, if x^* is an optimal solution of the given problem then there exists $\{\lambda^*, v^*\}$ such that:

$$\nabla f(x) + \lambda \nabla g(x) + v\nabla h(x) = 0 \quad \lambda g(x) = 0 \quad h(x) = 0$$

While $\lambda \ge 0$ and $g(x) \le 0$.

Solving the above system of equations yields points which are feasible for the minimum. The actual optimal points could be found by evaluating the objective function at each of these candidate points. However, if it's possible to prove that the objective function is a convex function of x then, the KKT conditions guarantee that the solution is optimal.

Index

A

Aamir, M., 85
Abadi, M., 179
Abbas, W., 101
access control (as security goal), 136
account enumeration attacks, 19–20, *20*, 21
account lockout settings, weaknesses, 21
active attacks, 100, 137, **137**; *see also specific types of attack*
active tags (in RFID), 2–5
Advanced Encryption Standard (AES), 23
aggregation, *see* data aggregation
agriculture applications, 103, **104**, 119, *172*
Alcaide, A., 68
ALERT (Anonymous Location-Based Efficient Routing Protocol), 146
Alexa (Amazon Echo devices), 151, 163–166, **165**; *see also* virtual assistants
Ali, S., 146
Aman, M. N., 72
Amazon Echo devices, 151, 163–166, **165**; *see also* virtual assistants
Amin, R., 73
Anderson, M., 126
Anonymous Location-Based Efficient Routing Protocol (ALERT), 146
Application Programming Interfaces (API), Open Source, 119
ARM mbed, **120**, 121
authentication
    biometric, *see* biometric authentication
    JSON Web Tokens (JWT), 116
    methods (overview), 68–73
    as security goal, 98, 136
authentication attacks, 99; *see also* man-in-the-middle (MITM) attacks; spoofing attacks; Sybil attacks
authentication authorization (in RIOT), 114
autoencoders, 176, *177*
    and dimensionality reduction, 176–180, *179–180*, *181–183*
availability (as security goal), 98, 136

availability violation attacks, *99*; *see also* battery drainage attacks; Denial of Service (DoS) attacks; Distributed Denial of Service (DDoS) attacks; jamming attacks; packet dropping attacks
Azimi, I., **104**

B

backdoors, 31
Bahga, A., 72
Balasubramanian, A., 68
Baldini, G., 126–127
Barreto, L., 70
battery drainage attacks, *99*, 100
biometric authentication
    and health monitoring systems, 68, 73, 74–77, **74**, **75**, *76*, 77; *see also* facial recognition (dimensionality reduction)
BipIO (API), 119
Bitcoin, 53, 55–56, 59
black hole attacks
    and IoV, **10**, 139, *140*, 145, **147**
    security solutions, 145, **147**
blockchain technology
    applications, 56–57, *56*
    benefits, 61–63
    Brooklyn Microgrid (BMG), 62–64, *64*
    challenges, 62
    decentralized authentication system, 72–73
    features, 55
    fundamental operations and structure, 53–55, *54*
    IoT network architecture, 57–59, *57*
    IoT network architecture with decentralized cloud, 59–61, *60*
    and machine learning, 102
    types, 55–56
bots and botnets (in DoS and DDoS attacks), 84
Brazil
    Civil Internet Framework, 151–152
    LGPD; *see also* General Law on Personal Data Protection (LGPD) (Brazil)

193